Nuclear Rules, Not Just Rights:
The NPT Reexamined

NUCLEAR RULES, NOT JUST RIGHTS:
THE NPT REEXAMINED

EDITED BY

HENRY D. SOKOLSKI

NONPROLIFERATION POLICY EDUCATION CENTER

Copyright © 2017 by Henry D. Sokolski
Nonproliferation Policy Education Center
Arlington, VA 22209
www.npolicy.org

Printed in the United States of America

All rights reserved. Except for brief quotations in a review, this book, or parts thereof, must not be reproduced in any form without permission in writing from the Nonproliferation Policy Education Center.

ISBN 978-0-9862-895-8-3

Contents

Introduction   1

1. What Does the History of the NPT Tell Us About Its Future?   5
Henry D. Sokolski

2. How We've Come to View the NPT: Three Pillars   37
Dean Rust

3. Early Thoughts on Safeguards and the NPT: Where Our Current Problems Began   102
Leonard Weiss

4. NPT's Naval Nuclear Propulsion Loophole   123
Jeffrey M. Kaplow

5. Another Gap in the NPT: How Israel and Others Get Outside Nuclear Help   154
Victor Gilinsky

6. Locking Down the NPT   165
Henry D. Sokolski and Victor Gilinsky

Appendix: Text of the Treaty on the Non-Proliferation of Nuclear Weapons   168

About the Contributors   177

Introduction

Nuclear Rules, Not Just Rights: The NPT Reexamined

Henry D. Sokolski

With 190 state members, the Nuclear Nonproliferation Treaty (NPT) is almost universal. However, it has fallen on hard times. North Korea violated it and withdrew in 2002. Israel, Pakistan, India, and North Korea—the nuclear-armed states most likely to use them—refuse to sign. Others—e.g., Syria, South Korea, and Egypt—have violated its safeguards and yet suffered no serious consequences. Also, with the Iran deal, enriching uranium or reprocessing spent reactor fuel, which can bring states to the very brink of bomb making, is now less taboo. Finally, with President Trump's suggestion that South Korea's and Japan's acquisition of nuclear weapons is inevitable, the prospect of the treaty lasting in perpetuity is easily open to question.[1]

It would be nice if it was otherwise. Without adherence to the treaty, a recent report speculated that Japan, South Korea, Poland, Turkey, and Saudi Arabia were all likely to acquire nuclear weapons before 2030 and that the UAE, Taiwan, Iran, Brazil, Nigeria, Argentina, Germany, Malaysia, Vietnam, and Egypt were also possibilities.[2]

---

1. Maggie Haberman and David E. Sanger, "Transcript: Donald Trump Expounds on His Foreign Policy Views," *The New York Times*, March 26, 2016, available from *http://www.nytimes.com/2016/03/27/us/politics/donald-trump-transcript.html*.

2. Clark Murdock and Thomas Karako, *Thinking about the Unthinkable in a*

Should major nuclear states, like the United States, conclude that making nuclear nonproliferation a priority is a mistake? Would it make more sense instead for Washington to bolster friendly states' nuclear capabilities to maintain advantage? Wasn't this precisely what the NPT was devised to prevent for fear that nuclear-armed states helping others get nuclear arms would beget more proliferation, thus making disarmament more remote and nuclear use more likely?

These questions would seem rhetorical especially if the NPT itself was a clear barrier to nuclear proliferation. But it is not. The NPT, even if it was adhered to, has serious loopholes and gaps. Some of these are well known. The NPT allows nuclear-armed states to base their weapons in NPT nonweapons states so long as the weapons remain under the "control" of the nuclear-armed donor state. The immediate concern here is what mischief this might allow for in the possible case of Pakistan or China basing nuclear weapons in other states, such as Saudi Arabia. This worry, however, has received so much attention that it's not immediately in prospect of being realized.

This brief volume focuses on six that are. The first of these concerns how the NPT is currently interpreted. Originally, Irish Foreign Minister Fred Aiken, the diplomat who first proposed the treaty, intended that the treaty be negotiated primarily to prevent the further spread of nuclear weapons. He explicitly subordinated the objectives of nuclear disarmament and of sharing civilian nuclear technology to the higher and, he argued, more pressing goal of nonproliferation. If nuclear weapons continued to spread, he reasoned, nuclear disarmament would be made practically impossible to negotiate. As for civilian nuclear power systems, he saw these rightly as bomb starter kits that required the most intense and innovative inspections to prevent from being diverted to make bombs.

Highly Proliferated World, Washington, DC: Center for Strategic and International Studies, 2016, available from *https://csis-prod.s3.amazonaws.com/s3fs-public/publication/160725_Murdock_ThinkingAboutUnthinkable_Web.pdf*.

Unfortunately, this is not how we now view the treaty. Instead, the United States and virtually all other countries, including Iran, argue that the NPT rests equally on three pillars or promises: Restraining nuclear nonproliferation and conducting inspections; providing the benefits of peaceful nuclear technology and recognizing the right to acquire them without discrimination; and limiting the nuclear arms race and encouraging nuclear and general disarmament. The problem with this formulation, however, is that it suggests one must promote disarmament and civilian nuclear energy exports as a quid pro quo for nuclear nonproliferation limitations and inspections. If so, the NPT risks becoming more a disarmament and civilian nuclear promotion agreement rather than a treaty to reduce the spread of nuclear weapons related technology.

Related is the question of just how effective the treaty's nuclear safeguards are. These are supposed to prevent military diversions not only from fresh and spent reactor fuel, but also from spent fuel reprocessing plants, plutonium and uranium fuel production plants, and uranium enrichment facilities. As our recent misgivings about Iran's enrichment activities and Japan's plans to reprocess spent fuel at Rokkasho suggest, though, detecting diversions from such activities to make bombs may not be all that reliable or early enough to assure they won't be built. The creators of the International Atomic Energy Agency (IAEA) knew that they did not know how to cope with these safeguards challenges; they hoped they might not have to face them or would be able to do so later when they might arise. They were wrong on both counts. The question now is will we admit as much and draw the line between what can and cannot be safeguarded more tightly?

Yet another concern is nuclear submarines. These require enriched uranium fuel. Under the NPT and the IAEA Statute, states that lack nuclear weapons can not only enrich uranium to power naval reactors, but remove the enriched uranium from IAEA safeguards as soon as this material is applied for an allowed nonexplosive military application (e.g., naval reactors). Brazil, Pakistan, and Iran all claim they have or are planning to have nuclear submarine programs. Each has had or is feared to have nuclear weapons aspirations.

A related problem is the NPT's inattention to assistance non-nuclear-weapon state parties to the treaty can give to other non-nuclear weapons state parties. This is allowed as is nuclear weapons delivery systems, nuclear weapons basing, and command and control assistance from both non-nuclear-weapon and nuclear weapon states to non-nuclear-weapons states. Germany, for example, could (and has) exported reactor and laser uranium enrichment technology to Iran. It also has exported advance air-independent propulsion submarines to Israel designed to accommodate the delivery of nuclear warheads. China and Russia have exported nuclear-capable missile technology to Iran. All of this is legal under the current interpretation of the NPT.

Finally, the NPT allows members to leave the treaty after 90 days of giving notice of cause to withdraw. This has allowed North Korea to acquire all that it needed to make nuclear weapons under the pretense of producing "peaceful" nuclear energy, violate its safeguard obligations under the NPT, set off a nuclear explosion, and withdraw.

All of these NPT loopholes are examined in this volume. In each chapter, each of these gaps are assessed along with what can be done to close them. One might argue that these critiques and suggestions come late, too late. But that is far from clear. The NPT is still in force and it is the only legal instrument there is to work these matters in a serious fashion.

CHAPTER 1

What Does the History of the NPT Tell Us About Its Future?[1]

Henry D. Sokolski

When experts discuss the prospects of the Nuclear Nonproliferation Treaty (NPT), they naturally focus on impending events. Will nonaligned nations tie their continued adherence to reaching a comprehensive test ban? Will Egypt, Algeria, Turkey, Saudi Arabia, and Iran live up to their NPT obligations? Will the NPT's inspectorate, the International Atomic Energy Agency (IAEA), strengthen its inspection procedures?

The answers to these questions—like the future itself, however—are necessarily speculative. In contrast, the NPT's history is known. More important, it is arguably more relevant to gauging the treaty's chances for future success. To understand the NPT's past, after all, is not only to understand what the treaty's original intentions were but to consider how practical and relevant these aims are today and how viable they are likely to be.

In general, of course, we already know what the NPT is supposed to do: Limit the spread of nuclear weapons. What we are less clear on, though, is exactly how the NPT is supposed to achieve this end. Was the goal of curbing the transfer of nuclear weapons technol-

---

1. This chapter was original published in Henry Sokolski, ed., *Fighting Proliferation: New Concerns for the Nineties*, Maxwell AFB, AL: Air University Press, 1996.

ogy to be subordinated to the NPT's stated aim of ending the arms race between Washington and Moscow? Did smaller nations, in fact, have a right—as the NPT's 10th article suggests—to withdraw from the treaty if, in their estimation, neither Washington nor Moscow had taken effective measures to end the nuclear arms race or if a neighboring adversary acquired nuclear weapons of its own? Did the NPT, in fact, reflect the view that nuclear proliferation was less of an evil than either of these outcomes?

What of nuclear safeguards? Were nuclear activities and materials that were quite close to bomb making or nuclear weapons themselves to be allowed if they were claimed to be for "peaceful purposes" and were acquired or transferred under international inspections? Did the drafters of the NPT's provisions for safeguards consciously limit the intrusiveness of inspections in order to protect any and all transfers of civilian nuclear energy?

Certainly, if the answer to these historical questions is yes, the NPT's future as an effective nonproliferation agreement would be in doubt. At a minimum, it would suggest that the prospects for strengthening the IAEA and NPT and for getting near-nuclear or undeclared-nuclear nations to join were distant.

The NPT's history, though, is not that clear. Certainly, it is true that the NPT's framers finally opposed intrusive IAEA inspections, encouraged the sharing of "peaceful" nuclear energy, described the greatest proliferation threat as being the superpowers' continued buildup of nuclear arms, and even claimed that nations had the right to acquire nuclear weapons under "extraordinary events." Yet, each of these propositions was debated and arguably balanced by the NPT's first two articles prohibiting the transfer or acquisition of nuclear weapons "directly or indirectly." These articles, first suggested by the Irish in 1958 as an intermediate step toward superpower nuclear arms control, presumed that the further spread of nuclear weapons threatened accidental and catalytic nuclear war and instability both for states with and those without nuclear weapons.

These "Irish" articles are important, then, if only because they seem at odds with the NPT's other provisions. These include language-backed by a substantial negotiating record that provides for NPT members' rights under Articles III, IV, and X to (1) withdraw from the treaty (and, thus, legally acquire nuclear weapons), (2) engage in the "fullest possible exchange" of nuclear technology, and (3) keep nuclear inspections under the NPT from "hampering [NPT members'] economic or technological development."

Critics of the NPT argue that Articles I and II should rule over the interpretation and implementation of the rest of the treaty.[2] However, this is neither the way the NPT is popularly understood nor the way most of the NPT's framers saw the treaty when they finalized it in 1968. Then, as now, the predominant nuclear threat in the eyes of the treaty's supporters was not accidental or catalytic war, but the possibility that nuclear competition between major nations might get out of hand, start a war, or—short of this—encourage nonweapons states to go nuclear. As the NPT's framers saw it, the best way to prevent this would be to agree to total nuclear disarmament, while mutual nuclear deterrence at very low levels of nuclear armament among nations would be second best. Indeed, smaller nations might prefer to acquire their own nuclear forces rather than allow an ever-escalating and threatening nuclear arms race between the major nuclear states go unchallenged or have to depend on unreliable superpower guarantees of nuclear security alliance.

From this perspective, asking states without nuclear weapons to forgo acquiring nuclear weapons is asking them to forgo exercising a "right" that could be in their national security interest. As such, forswearing nuclear weapons required a quid pro quo: Encourage the superpowers to take "effective measures" to end the nuclear arms race and facilitate the fullest possible transfer of civilian nuclear technology (which the nuclear powers gained by developing

---

2. See, for example, Frank Barnaby and Shaun Burnie, *The Nonproliferation Treaty: A Critical Assessment,* Amsterdam, the Netherlands: Greenpeace International, January 17, 1994.

weapons) from the nuclear haves to the nuclear have-nots.

Such deal making, however, is unnecessary if one focuses on the security concerns highlighted in the Irish's original United Nations (UN) resolution of 1958. Curbing the threat of accidental and catalytic nuclear war would be a good that both states with and without nuclear weapons would benefit from—a good worth subordinating all other aspects of the NPT to achieve and having effective (and even discriminatory) safeguards to achieve.

This, then, is the challenge facing today's supporters of the NPT. They must recognize that there are two different ways to interpret the treaty: Through the lens of the Irish resolutions (i.e., Articles I, II, and—arguably—III) or through the articles that follow. For the policymaker, making this choice is critical to determining just how viable the NPT is likely to be and what, if anything, remains to be done.

To choose wisely requires an understanding of what sort of proliferation threat the NPT was originally intended to address; how and why this original concern was largely displaced by the new concerns noted above; how much of a tension between these views remained at the time of the NPT's signing in 1968; and which of these views makes more sense today. In short, we must go back to the NPT's origins.

1958-65: The Irish Resolution and Preventing Catalytic Nuclear War and the Further Spread of Nuclear Weapons

Although the proliferation of nuclear weapons is now synonymous with the spread of know-how, nuclear materials, and specialized equipment to rogue states such as Iran, this was not the central worry animating those who first suggested the need for an international nonproliferation agreement in the late 1950s. Instead, their concern was the actual and proposed American transfers of nuclear weapons to Germany and the North Atlantic Treaty Organization (NATO).

Starting with President Dwight Eisenhower's Administration in 1953, the United States began to deploy nuclear artillery in Europe for use by NATO forces under a "dual key" control arrangement. The United States had custody of the nuclear-artillery warheads, while U.S. and NATO armies had nuclear-capable artillery tubes integrated into their ground forces. If an occasion arose when the U.S. president deemed use of the nuclear artillery necessary, he could order the release of the nuclear warheads to the NATO commander, and the commander of the NATO ally would give authority to release use of the nuclear-capable artillery tubes. Following this model, the United States was able to deploy nuclear weapons not only to NATO ground forces but to U.S. and allied air forces in Europe without losing control of the weapons themselves.

Unfortunately, Warsaw Pact members and the world's neutral powers believed that U.S. authority over these weapons was less than complete. In 1956 and 1957, the Soviet Union was so concerned about the United States' stationing of nuclear weapons in Germany that it proposed a ban on the employment of nuclear weapons of any sort in Central Europe.[3] The United States, meanwhile, submitted a draft disarmament plan before the UN Disarmament Commission in which transfer of control of U.S. nuclear weapons to NATO allies was permitted if their use was necessary to fend off an armed attack.[4]

In 1958 concern with controls over such nuclear transfers was heightened further when the U.S. Congress passed an amendment to the U.S. Atomic Energy Act that permitted the transfer of weapons

---

3. See "Soviet Proposal Introduced in the Disarmament Subcommittee: Reduction of Armaments and Armed Forces and the Prohibition of Atomic and Hydrogen Weapons, May 18, 1957," in U.S. Department of State, *Documents on Disarmament, 1945-1959*, Washington, DC: Government Printing Office, 1960, pp. 756-57.

4. Ibid. See "Western Working Paper Submitted to the Disarmament Subcommittee: Proposals for Partial Measures of Disarmament, August 29, 1957," in *Documents on Disarmament, 1945-1959*, p. 870.

materials, design information, and parts to nations that had "made substantial progress in the development of nuclear weapons."[5] Also, with the continued transfer of nuclear weapons to NATO, U.S. control arrangements became less rigid: One congressional investigation discovered German aircraft that were fueled, ready to take off at a moment's notice, and loaded with U.S. nuclear weapons.[6]

This trend toward laxer U.S. restraints on authority for the transfer of nuclear weapons came at the same time as progress toward disarmament negotiations in the UN had reached an impasse. The United States and the Soviet Union had agreed to a voluntary moratorium on nuclear testing in the fall of 1958, but the United States and its allies tied their continued adherence to this test ban to progress toward disarmament and a general easing of tensions. Last, but hardly least, the United States had threatened or considered using nuclear weapons on at least six separate occasions since Eisenhower had assumed the presidency in 1953.[7]

Against this backdrop, the Irish offered their draft resolution concerning the "Further Dissemination of Nuclear Weapons" before the First Committee of the General Assembly of the UN on October 17, 1958. This resolution was quite modest, recognizing that "an increase in the number of states possessing nuclear weapons may occur, aggravating international tensions" and making disarmament

---

5. The Atomic Energy Act of 1954, U.S. Code sees. 54, 64, 82, 91(c), 92 as amended (1954).

6. See George Bunn, *Arms Control by Committee: Managing Negotiations with the Russians*, Stanford, CA: Stanford University Press, 1992, p. 62.

7. The Eisenhower administration had threatened to use or considered using nuclear weapons to end the Korean War in 1953, to save the French in Vietnam in 1954, to save the Republic of China in 1954, 1955, and 1958, and to prevent any invasion of Kuwait in 1958. Atomic howitzers also were deployed by U.S. forces landing in Lebanon in 1958. The Russians, meanwhile, threatened the use of nuclear weapons to end the Suez crisis in 1956. See Peter Lyon, *Eisenhower: Portrait of the Hero*, Boston: Little, Brown and Company, 1969, pp. 534, 541, 583, 606, 610, 624, 639-40, 719, 775-76, 784.

"More difficult." It went on to recommend that the General Assembly establish an ad hoc committee to study the dangers inherent in the further dissemination of nuclear weapons.

The Irish offered to amend the resolution to urge parties to The UN's disarmament talks not to furnish nuclear weapons to any other nation while the negotiations were under way and to encourage other states to refrain from trying to manufacture nuclear weapons, but Western support for the amendment was thin. On October 31, 1958, the Irish withheld the resolution when it became clear that no NATO nation was yet ready to endorse the initiative.[8]

The Irish, however, pursued the idea. The following year their foreign minister resubmitted yet another version of the resolution to the General Assembly and made it clear that the proposal was a minimal proposition which all parties ought to accept. It was "hardly realistic," he argued, to expect any "early agreement on the abolition of nuclear weapons." But "what we can do," he argued, "is to reduce the risks which the spread of these weapons involves for this generation, and not to hand on to our children a problem even more difficult to solve than that with which we are now confronted." Indeed, the Irish foreign minister argued that "if no such agreement is made, they [the nuclear powers] may well be forced by mutual fear and the pressure of their allies, to distribute these weapons, and so increase geometrically the danger of nuclear war."[9]

Why was such nuclear proliferation so dangerous and likely? First, without an international nonproliferation agreement, "a sort of atomic *sauve-qui-peut*" was likely in which states, "despairing of

---

8. The resolution initially passed with 37 affirmative votes, but 44 nations—including the United States, United Kingdom, Italy, Japan, France, Greece, Belgium, Turkey, and the Netherlands—abstained. See "Irish Draft Resolution Introduced in the First Committee of the General Assembly: Further Dissemination of Nuclear Weapons, October 17, 1958," in *Documents on Disarmament, 1945-1959*, pp. 1185-86.

9. See "Address by the Irish Foreign Minister [Aiken] to the General Assembly, November, September 23, 1959 [extract]," in ibid., pp. 1474-78.

safety through collective action," would seek safety for themselves by getting nuclear weapons of their own.[10]

This trend was likely to get worse, the Irish argued, since there was "no conceivable addition" to the list of countries possessing nuclear weapons which would not cause a change in the pattern of regional and world politics that could be "great enough to destroy the balance of destructive weapons ... which has given the world the uneasy peace of the last few years."[11] As the Irish foreign minister later explained,

> the sudden appearance of nuclear weapons and their almost instantaneous long-range delivery systems in a previous non-nuclear State may be tantamount, in the circumstances of the world today, to pushing a gun through a neighbor's window. . . . It may even be regarded as an act of war by neighboring countries who have not the second strike nuclear capacity possessed by great nuclear Powers....[who] may be able to eliminate the threat by taking limited measures.[12]

Second, faced with these threats, nations without nuclear weapons would try to acquire them from their nuclear-armed allies, who, out of a misguided sense of political convenience, were likely to be cooperative. All this would do, however, is give these smaller nations "the power to start a nuclear war, or to engage in nuclear blackmail—conceivably against a former ally." In short, without an international agreement against further transfers of nuclear weapons, accidental and catalytic wars would become more likely, and na-

---

10. Ibid.

11. See "Statement by the Irish Foreign Minister [Aiken] to the First Committee of the General Assembly, November 13, 1959," in ibid., pp. 1520-26.

12. See "Statement of Irish Foreign Minister [Aiken] to the First Committee of the General Assembly, November 6, 1962," in U.S. Arms Control and Disarmament Agency, *Documents on Disarmament, 1962*, Washington, DC: Government Printing Office, 1963, pp. 1025-28.

tions would drift into "a nightmare region in which man's powers of destruction are constantly increasing and his control over these powers is constantly diminishing."[13]

Finally, nuclear weapons technology itself was becoming more available. As the Irish foreign minister explained, weapons-usable plutonium was a direct by-product of nuclear electrical-power reactors, and these generators were being built in states without nuclear weapons. It would become increasingly difficult, he believed, for the governments of these countries to "resist domestic pressure to take the further step of producing nuclear weapons [on the] grounds of economy and security, if not for considerations of prestige."[14]

These considerations were all factored into the original bargain inherent in the Irish resolution. The states with nuclear weapons would forgo relinquishing control of their weapons to their allies, and the states without nuclear weapons would refrain from manufacturing or acquiring them and accept inspection of their "reactors and territories" to ensure that they were living up to their undertakings. This was the full extent of the bargain. All states—with or without nuclear weapons—would be better off because the possibility of accidental or catalytic war would be reduced. Beyond this, nonweapons states would be spared the expense of having to develop strategic weapons, and the weapons states would have less reason to advance the qualitative development of their own strategic systems.

---

13. See "Statement by the Irish Foreign Minister, November 13, 1959," in *Documents on Disarmament, 1945-1959*, pp. 1520-26. In this speech, Foreign Minister Aiken attributes these views to Howard Simons, "World-Wide Capabilities for Production and Control of Nuclear Weapons," *Daedalus*, Vol. 88, No. 3, Summer 1959, pp. 385-409, which was a summary of "The Nth Country Problem: A World-Wide Survey of Nuclear Weapons Capabilities," a study by the American Academy of Arts and Sciences, which would be published by the National Planning Association in 1959.

14. "Statement by the Irish Foreign Minister, November 13, 1959," in *Documents on Disarmament, 1945-1959*, pp. 1520-26.

The Irish insisted on no direct linkage with progress on capping or reversing the arms rivalry between Moscow and Washington. Nor was there any notion that the nuclear nations should offer "peaceful" nuclear technology to the nonweapons states to get them to open their territories to inspection. In fact, as the Irish foreign minister later made clear, nonweapons nations ought to welcome having their nuclear facilities inspected or, at least, not object since they might later serve as arms control test beds. Nor, he argued, should the inequality of nonweapons states opening their nuclear facilities to inspections (from which nuclear states would be exempt) be seen as involving any "loss of prestige." After all, several nonweapons states had already endorsed the idea of regional disarmament and European nuclear-weapons free zones that required asymmetrical inspections. Nonproliferation inspections were only an extension of the same idea.[15]

The United States and other states with nuclear weapons, however, initially had misgivings about the Irish resolutions. As has already been noted, most NATO nations abstained when the Irish resolution was first put to a vote in 1958. In 1959, though, the Soviet Union also opposed the resolution, complaining that it was too permissive: It would allow the Americans to transfer nuclear weapons to European soil so long as the United States "retained control" of the weapons. Meanwhile, France abstained, arguing that the transfer of fissionable materials and nuclear weapons was difficult to control and that the real problem was ending manufacture of these items. At the time, France was itself getting ready to test its first nuclear weapon and was assisting the Israelis in their nuclear weapons efforts.[16]

As for the United States, it supported the 1959 Irish resolution after

---

15. See "Statement of Irish Foreign Minister ... November 6, 1962," in *Documents on Disarmament, 1962*, pp. 1025-28.

16. See Lawrence Scheinman, *Atomic Energy Policy in France under the Fourth Republic*, Princeton, N.J.: Princeton University Press, 1965, pp. 183 ff. and Avner Cohen, "Stumbling into Opacity: The United States, Israel, and the Atom, 1960-63," *Security Studies*, Vol. 4, No. 2, Winter 1994, pp. 199-200.

abstaining in 1958, arguing that it permitted serious study of critical issues. Yet, when the resolution was modified in 1960 to call upon the weapons states to declare at once their intention to "refrain from relinquishing control of such weapons to any nation not possessing them and from transmitting to it the information necessary for their manufacture," the United States again objected. Although the Soviets decided to reverse themselves and support the draft, the United States at the time was pushing the idea of giving NATO nuclear submarines and missile boats for a multilateral force (MLF). As such, the U.S. representative to the UN complained that the resolution failed to recognize the critical responsibility of the nations with nuclear weapons. The U.S. representative went on to ask how the Irish could expect other nations to forgo nuclear weapons if the weapons states refused to end their own nuclear buildup. Besides, he argued, a commitment of indefinite duration of the sort the resolution called for was unverifiable.[17]

The United States again objected in 1961, when the Swedes resubmitted a similar resolution recommending that

> an inquiry be made into the conditions under which countries not possessing nuclear weapons might be willing to enter into specific undertakings to refrain from manufacturing or otherwise acquiring such weapons and to refuse to receive, in the future, nuclear weapons in their territories on behalf of any other country.[18]

The resolution's new language worried the United States. The resolution was no longer focused on restraining weapons nations from "relinquishing control" of nuclear weapons but on getting

---

17. See United Nations Department of Political and Security Council Affairs, *The United Nations and Disarmament, 1945-1970*, New York: United Nations Publications, 1971, pp. 260-61.

18. The Swedes submitted this resolution, 1664 (16), December 4, 1961. See ibid., p. 265.

nonweapons nations to refuse receiving nuclear weapons in their territories. In short, it appealed to all of NATO to stop hosting U.S. nuclear weapons. This point was hardly lost on the Soviets, who immediately incorporated the Swedish language (i.e., "refrain from transferring control [and] refuse to admit the nuclear weapons of any other states into their territories") into their own draft treaty for general and complete disarmament in 1962.[19]

The United States objected to the Swedish resolution, complaining that it effectively called "into question the right of free nations to join together in collective self-defense, including the right of self-defense with nuclear weapons if need be." Yet, the U.S. representative was equally insistent that the United States supported the goal of nonproliferation. His proof was that the United States draft program for general and complete disarmament—like the Irish resolution—required states with nuclear weapons to "refrain from relinquishing *control*" (emphasis added) of nuclear weapons to nonweapons states.[20]

1965-68: Bargaining to Keep States from Exercising their "Right" to Acquire Nuclear Weapons

For the next four years, the United States continued to insist that it was interested in promoting nuclear nonproliferation.[21] However,

---

19. Ibid.

20. See "Statement by the United States Representative [Yost] to the First Committee of the General Assembly: Spread of Nuclear Weapons, November 30, 1961," in U.S. Arms Control and Disarmament Agency, *Documents on Disarmament, 1961*, Washington, DC: Government Printing Office, 1962, pp. 691-92.

21. See, for example, "Statement by ACDA Director Foster to the Eighteen Nation Disarmament Committee: Nondissemination of Nuclear Weapons, February 6, 1964," in U.S. Arms Control and Disarmament Agency, *Documents on Disarmament, 1964*, Washington, DC: Government Printing Office, 1965, pp. 32-33, in which restraint in international nuclear nonproliferation was urged since without it there "would be no rest for anyone ... no stability, no real security and no chance

it opposed a variety of nonproliferation resolutions backed by the Soviets, Swedes, and others, which, if accepted, would have jeopardized existing nuclear-sharing arrangements with NATO on the possibility of creating a multilateral nuclear force for a "United States of Europe." Ultimately, the United States focused on reaching an international nuclear nonproliferation agreement only when it became clear that Germany and other NATO nations were not keen on reaching an MLF agreement. With the MLF disposed of and the Soviets willing to accept language that would allow the Americans to deploy nuclear weapons in NATO—assuming they were kept under U.S. control—the United States was ready to negotiate a nonproliferation agreement.[22]

By early 1966, though, the terms of UN debate over proliferation had changed. Whereas in 1958, nonproliferation was seen as a good in itself—equally beneficial to states with and without weapons—by the early 1960s, smaller nations perceived nuclear nonproliferation as a potential obstacle to assuring their national security, while the United States and Soviet Union continued to refine and expand their own nuclear arsenals.

Another key difference in the debate was how nations viewed superpower nuclear deterrence. In 1959, the Irish downplayed the threat presented by nuclear superpower rivalry: "That situation, fraught with danger as it is, is nonetheless one with which we have managed to live for a number of years. Techniques have been evolved to deal with it." The key concern wasn't with this set of dangers but with those "likely to flow with the wider dissemination of nuclear weapons."[23]

---

of effective disarmament." It was also argued that because the acquisition of nuclear weapons by smaller countries would "increase the likelihood of the great Powers becoming involved in what would otherwise remain local conflicts," both the security of weapons and nonweapons states in U.S. eyes was at stake.

22. See Bunn, pp. 66-75.

23. See "Statement by the Irish Foreign Minister, November 13, 1959," in *Documents on Disarmament, 1945-1959*, pp. 1520-26.

By the mid-1960s, however, faith in the stability of the superpower nuclear "balance" and concerns about the threat of accidental and catalytic war had begun to wane. In their place, worries about the superpower "arms races" and the threat of the superpowers' "nuclear imperialism" over nonnuclear nations gained popularity. As India's UN representative explained in 1966,

> [the] dangers of dissemination and independent manufacture [of nuclear weapons] pale into the background when one views the calamitous dangers of the arms race which is developing today as a result of the proliferation of nuclear weapons by the nuclear weapon Powers themselves, large and small. For many years now, the superpowers have possessed an over-kill or multiple-destruction capacity and even their second-strike capabilities are sufficient to destroy the entire world. They have hundreds of missiles of varying ranges which are capable of devastating the surface of the earth. They are continuing to test underground, miniaturizing warheads, improving penetration capabilities and sophisticating their weapons and missiles. The other nuclear weapons powers are also following the same menacing path, conducting atmospheric weapons tests, and submarines. Only four days ago, the People's Republic of China conducted yet another weapons test, firing an intermediate-range guided missile with a nuclear warhead. When we talk of the dangers of the arms race, therefore, we face the dangers of the most titanic proportions. It is here that the proliferation of nuclear weapons has its most catastrophic consequences.[24]

---

24. See "Statement by the Indian Representative [Trivedi] to the First Committee of the General Assembly: Nonproliferation of Nuclear Weapons, October 31, 1966," in U.S. Arms Control and Disarmament Agency, *Documents on Disarmament, 1966*, Washington, DC: Government Printing Office, 1967, pp. 679.

Egypt's representative to the UN disarmament talks made the same point somewhat differently:

> The nonnuclear countries will in law renounce their right to nuclear weapons, but nuclear stockpiles and the threat of a nuclear confrontation will in fact continue to exist indefinitely....This de facto situation could always constitute an incitement to manufacture or acquire nuclear weapons. To diminish this risk still further it will be necessary, pending the complete elimination by radical measures of nuclear stockpiles and the nuclear threat, to include in the treaty a formal and definite indication of what the nuclear Powers propose to do with the existing nuclear armament.[25]

Why did this shift occur? First, non-nuclear nations who were eager for a nonproliferation treaty in the very early 1960s but frustrated by the impasse created by the Soviet Union, United States, and NATO nations over the issue decided to work without the superpowers' cooperation. As has already been noted, in 1961 the Swedes submitted a resolution before the UN General Assembly calling for an inquiry as to the conditions under which nonweapons states might be willing to refrain from acquiring nuclear weapons. The idea here was to force the nuclear states' hand by demonstrating the popularity of nuclear nonproliferation and threatening to promote it without the superpowers. However, the very premise of the inquiry—that nonweapons nations would naturally acquire nuclear weapons unless certain "Conditions" were met—was at odds with the idea that nonproliferation was equally a security imperative for both weapons and nonweapons states.

Second, beginning in the late 1950s, an intellectual shift occurred in the way nuclear arms and deterrence were viewed. During this

---

25. See "Statement of the Egyptian Representative [Khallaf] to the Eighteen Nation Committee on Disarmament, March 3, 1966," in ibid., pp. 156-57.

period, a new nuclear theory—finite deterrence—emerged. According to this view, smaller nations could keep larger nuclear powers from threatening them militarily by acquiring a small number of nuclear weapons of their own. With their limited nuclear arsenal, the smaller nations might not be able to prevail in war against a larger power but could effectively "tear an arm off" by targeting the larger nation's key cities and thus deter such nations from ever attacking.[26] Closely related to this point was a critique of the superpowers' constant quantitative and qualitative improvement of their strategic forces. This buildup was considered unnecessary and provocative because a nation needed only a small nuclear arsenal to threaten to knock out an opponent's major cities.[27]

In 1962 this view was reflected in replies to the UN secretary-general's inquiry about the conditions under which nonweapons states "might be willing to enter into specific undertakings to refrain" from acquiring weapons. 62 nations replied, most of them wanting specific neighbors or all the states within their region to forswear acquiring nuclear weapons as a condition for their doing likewise. Other nations, such as Italy, wanted the nuclear powers to halt their nuclear buildup.[28] Meanwhile, the three nuclear powers that answered the

---

26. One of the earliest expressions of this idea can be found in Jacob Viner, "The Implications of the Atomic Bomb for International Relations," in *International Economics: Studies by Jacob Viner*, Glenco, Ill.: Free Press, 1951, pp. 300-309. For the earliest popular presentation of finite deterrence theory, see Pierre M. Gallois, "Nuclear Aggression and National Suicide," *The Reporter*, November 18, 1958, pp. 22-26.

27. See, for example, P. H. Backus, "Finite Deterrence, Controlled Retaliation," *U.S. Naval Institute Proceedings*, March 1959, pp. 23-29 and George W. Rathjens, Jr., "Deterrence and Defense," *Bulletin of the Atomic Scientists*, September 1958, pp. 225-28.

28. In fact, Italy first voiced reservations about agreeing not to acquire nuclear weapons unless the nuclear weapons nations promised to disarm in a NATO gathering held in February of 1962. Later that year, however, it acquiesced and supported a U.S. draft resolution that would allow the use of U.S. weapons by a multilateral NATO naval force. For details, see George Bunn, Roland M. Timerbaev, and James F. Leonard, "Nuclear Disarmament: How Much Have the Five Nuclear

inquiry indicated that general and complete disarmament was the best solution.[29]

For the next two years, the debate over the merits of establishing a European MLF made it impossible for the Soviet Union, United States, and most NATO nations to reach any agreement over nuclear nonproliferation.[30] At the very least, no progress in nonproliferation seemed likely until moves toward disarmament made progress. The world's nonaligned nonweapons states, on the other hand, were eager to secure a separate nonproliferation treaty and called on the UN to convene an international conference to negotiate such an agreement.[31] In June of 1965, India and Sweden suggested a new approach to the UN Disarmament Commission: A nonproliferation agreement combined with measures that would begin to cap the arms race between the superpowers. Italy also suggested imposing a time limit on the non-nuclear nations' agreement to refrain from acquiring nuclear weapons. Advocates of this limit—a threat of coercive leverage *in potentia*—argued that it would serve as an "inducement" to the superpowers to disarm. With support from the world's nonaligned nations, the resolution passed overwhelmingly.[32]

Powers Promised in the Non-Proliferation Treaty?" in John B. Rhinelander and Adam M. Scheinman, eds., *At the Nuclear Crossroads: Choices about Nuclear Weapons and Extension of the Non-Proliferation Treaty*, Lanham, Md.: University Press of America, Inc., 1995, p. 15.

29. See *The United Nations and Disarmament, 1945-1970*, p. 266.

30. See, for example, the exchange between the Soviet and U.S. representatives to the Eighteen Nation Disarmament Committee, July 2, 1964, in *Documents on Disarmament, 1964*, pp. 241-56.

31. For a review of the nonaligned nations' actions along these lines, see "Statement by the U.A.R. Representative [Fahmy] to the First Committee of the General Assembly: Nonproliferation of Nuclear Weapons, October 22, 1965," in ibid., pp. 485-90.

32. See *The United Nations and Disarmament, 1945-1970*, p. 269. Italy and others continued to promote this idea through 1967. See, for example, "Statement by the Burmese Representative [Maung Maung] to the Eighteen Nation Disar-

From this point on, the debate over reaching a nuclear nonproliferation agreement presumed that nonweapons nations had a right to acquire nuclear weapons and that the only question was what they should get in exchange for not exercising it. Each nation expressed this right in a different fashion. For China, it was essential that non-nuclear nations not be "deprived of their freedom to develop nuclear weapons to resist U.S.-Soviet nuclear threats."[33]

For Brazil, the prerogative of non-nuclear nations to go nuclear was nothing less than their right to self-defense. As Brazil's representative explained,

> if a country renounces the procurement or production by its own national means of effective deterrents against nuclear attack or the threat thereof, it must be assured that renunciation–a step taken because of higher considerations of the interests of mankind–will not entail irreparable danger to its own people. The public could never be made to understand why a government, in forswearing its defense capability, had not at the same time provided reasonable and lasting assurances that the nation would not be, directly or indirectly, the object of total destruction or of nuclear blackmail.[34]

For Brazilians this meant that any nuclear nonproliferation agree-

---

mament Committee: Nonproliferation of Nuclear Weapons, October 10, 1967"; and "Statement by the Italian Representative [Caracciolo] to Eighteen Nation Disarmament Committee: Draft Nonproliferation Treaty, October 24, 1967," in U.S. Arms Control and Disarmament Agency, *Documents on Disarmament, 1967*, Washington, DC: Government Printing Office, 1968, pp. 463 and 529.

33. See, for example, "Chinese Communist Comment on Draft Nonproliferation Treaty, September 3, 1967," in *Documents on Disarmament, 1967*, p. 381.

34. See "Statement by the Brazilian Representative [Azeredo da Silveira] to the Eighteen Nation Disarmament Committee: Draft Nonproliferation Treaty, August 31, 1967," in ibid., p. 370.

ment had to include guarantees that states with nuclear weapons would not use or threaten to use them against states without such weapons.

Other states, however, thought that nothing less than nuclear disarmament was necessary. Tunisia, like Brazil, was "not happy about renouncing [its] right to acquire nuclear weapons" but thought that it was too poor ever to try to acquire them and thus could be truly secure only in a disarmed world.[35] Sweden, which was still developing a nuclear weapons option of its own,[36] shared Tunisia's views but saw giving up "the most powerful weaponry that has ever been produced by man" as something it—as one of the "smaller and more defenseless nations"—could do only if the superpowers disarmed.[37]

India, which was also developing a nuclear weapons option,[38] was the most outspoken in defending its "right" to "unrestricted" development of nuclear energy. This stance, in part, was simply a reflection of India's established opposition to international safeguards, which—it had argued since the early 1950s—would interfere with its economy's development and its "inalienable right [to] produce and hold the fissionable material required for [its] peaceful power programs."[39] After China exploded its first nuclear device in May

---

35. See "Address by President Bourguiba of Tunisia to the General Assembly, September 27, 1967 [extract]," in ibid., p. 429.

36. See Steve Coil, "Neutral Sweden Quietly Keeps Nuclear Option Open," *The Washington Post*, November 25, 1994, p. A1.

37. See "Statement by the Swedish Representative [Alva Myrdal] to the Eighteen Nation Disarmament Committee: Nonproliferation of Nuclear Weapons, October 3, 1967," in *Documents on Disarmament, 1967*, p. 444.

38. For a brief history of India's nuclear weapons program, see Leonard Spector, *Nuclear Proliferation Today*, New York: Vintage Books, 1984, p. 23 ff.

39. See Roberta Wohlstetter, *The Buddha Smiles: Absent-minded Peaceful Aid and the Indian Bomb*, Energy Research and Development Administration, Monograph 3, contract no. (49-1)-3747, Marina del Rey, CA: Pan Heuristics,

of 1964, though, protecting this right became even more imperative. As the Indian minister of external affairs explained in 1967,

> most of the countries represented at the disarmament committee appreciated India's peculiar position with regard to the nonproliferation treaty.... China would be a nuclear state which would not be called upon to undertake any obligations. India could have become a nuclear country if it had exploded the bomb as China did. But because India had shown restraint, a desire for peace, and opposition to the spread of nuclear armaments, under this treaty it would find itself in a much worse position than China .... The result of our restraint is that we are a nonnuclear power which will have to suffer all the disadvantages. On the other hand, China, which has shown no restraint, will not suffer from any disadvantage even if it signs the treaty, as it is already a nuclear power.[40]

What were the Indians talking about? The minister of external affairs left little doubt that they were referring to every nuclear "advantage" the weapons nations enjoyed—including nuclear testing. After all, he noted, the draft nonproliferation treaty would "seriously hamper and impede" peaceful nuclear research since it would prevent nonnuclear countries from undertaking underground explosions for the purpose of carrying out nuclear research while imposing no such obligation on states with nuclear weapons.[41] The ability to produce weapons-usable materials free from intrusive and discriminatory international safeguards and the freedom to develop all aspects of nuclear energy—including nuclear explosives, the minister contin-

---

April 30, 1977, pp. 30-75.

40. See "Extract from News Conference Remarks by the Indian External Affairs Minister [Chagla], April 27, 1967," in *Documents on Disarmament 1967*, pp. 204-205.

41. Ibid.

ued—was critical to secure India's "sovereign right of unrestricted development" of nuclear energy.[42]

If it were just India making these arguments, they might be dismissed as being peculiar to a nation "exposed to nuclear blackmail."[43] Yet, Brazil's representative shared India's views, arguing that

> nuclear energy plays a decisive role in [the] mobilization of resources. We must develop and utilize it in every form, including the explosives that make possible not only great civil engineering projects but also an ever-increasing variety of applications that may prove essential to speed up the progress of our peoples. To accept the self-limitation requested from us in order to secure the monopoly of the present nuclear-weapon powers would amount to renouncing in advance boundless prospects in the field of peaceful activities.[44]

At the time, Brazil was developing a nuclear weapons option of its own.[45]

It would be wrong, however, to dismiss Brazil's and India's interest in peaceful nuclear explosives (PNE) and sensitive nuclear activities as a cynical move. The United States, after all, had been touting the possible advantages of PNEs since the early 1960s as why it opposed reaching a comprehensive nuclear test ban with the

---

42. See "Statement by the Indian Representative [Trivedi] to the Eighteen Nation Disarmament Committee: Nonproliferation of Nuclear Weapons, May 23, 1967," in ibid., p. 235.

43. Ibid.

44. See ibid. and "Statement by the Brazilian Representative [Correa da Costa] to the Eighteen Nation Disarmament Committee: Peaceful Uses of Nuclear Energy, May 18, 1967," in ibid., p. 226.

45. For a description of Brazil's attempt to secure a safeguarded military production reactor during this period, see Spector, pp. 236-38.

Soviets. The United States also was enthusiastic about the need to develop fast-breeder reactors that would use reprocessed plutonium fuels.[46] Thus, Nigeria, Mexico, and Ethiopia, who had no nuclear programs, were every bit as insistent as India and Brazil that any treaty on nonproliferation not place them "in a position of perpetual inferiority in any field of knowledge."[47] Nigeria's recommendation to solve this problem was

> that non-nuclear weapons powers would not only have nuclear explosives, through an international organization, for their peaceful projects but also have opportunities for their scientists to develop to the full their intellectual capabilities in all fields, including that of nuclear-explosive technology.[48]

These nations were just as adamant that whatever international safeguards the NPT required not interfere with their development of new power reactors and fuels. In this, they were joined by Japan and Germany, who feared that the United States and Soviet Union would use the NPT's safeguard provisions to steal industrial nuclear secrets from their civil nuclear programs. As Germany's foreign

---

46. See Albert Wohlstetter et al, *Swords from Plowshares: The Military Potential of Civilian Nuclear Energy*, Chicago: University of Chicago Press, 1979, pp. 85-86; and idem, *Can We Make Nuclear Power Compatible with Limiting the Spread of Nuclear Weapons?* Vol. 1-1, *The Spread of Nuclear Bombs: Predictions, Premises, Policies*, ERDA contract no. E(49-1)-3747, Los Angeles: Pan Heuristics, November 15, 1976, pp. 9-32, 89-108.

47. See, for example, "Statement by the Ethiopian Representative [Zelleke] to the Eighteen Nation Disarmament Committee: Nonproliferation of Nuclear Weapons, October 5, 1967" and "Statement by the Mexican Representative [Castaneda] to the Eighteen Nation Disarmament Committee: Latin American Nuclear-Free Zone, May 18, 1967," in *Documents on Disarmament, 1967*, pp. 228, 449-50.

48. See "Statement by the Nigerian Representative [Sule Kolo] to the Eighteen Nation Disarmament Committee: Draft Nonproliferation Treaty, August 31, 1967," in ibid., p. 377. The Germans also shared this view. See, for example, "Statement by Foreign Minister Brandt to the Bundestag: Nonproliferation of Nuclear Weapons, February 1. 1967 [extracts]," in ibid.

minister explained in 1967,

> The unhindered civilian utilization of the atom is a vital interest of the Federal Republic.... It is known that German scientists are working with the prospect of success on the development of the second generation of reactors, the so-called fast breeders.... We go on the assumption that the placing into effect of controls does not interfere with the economic operations of factories, does not lead to the loss of production secrets, but counters the dangers of misuse. For this purpose it is adequate to control the end-product points, and to have a control which possibly could be exercised by automated instruments.[49]

Germany's foreign minister argued that nations like his own were already apprehensive of states with nuclear weapons trying to monopolize the civilian nuclear field by dint of their commanding lead in military nuclear technology.[50] At least as great a worry, he argued, was the extent to which inspections under the proposed NPT might compromise the pace and commercial confidentiality of civil nuclear developments by nonweapons states.

In the end, the NPTs preamble and Article III stipulated that nations like Germany could meet their safeguards obligations through somewhat less threatening but "equivalent" procedures under EURATOM (Western Europe's nuclear safeguarding organization), that inspections would be restricted to monitoring the flows of source and fissionable materials at "certain strategic points," and that they would be designed "to avoid hampering the economic or technological development of the Parties."

The NPT also emphasized in Articles IV and V that nothing in the

---

49. See "Statement by Foreign Minister Brandt to the Bundestag on Proposed Nonproliferation Treaty, April 27, 1967," in ibid., pp. 211-12.

50. Ibid.

treaty should be "interpreted as affecting the inalienable right of all the Parties to the Treaty to develop research, production and use of nuclear energy for peaceful purposes without discrimination." Indeed, the treaty called on all parties to "undertake to facilitate [the] fullest possible exchange of equipment, materials and technological information for the peaceful uses of nuclear energy." The treaty established procedures for sharing the benefits of peaceful nuclear explosives, although it prohibited the direct transfer of explosive devices to or development by nonweapons states.

Finally, in Article VI the treaty called on the weapons states to "pursue negotiations in good faith on effective measures relating to the cessation of the nuclear arms race at an early date and to nuclear disarmament." Even the Italians' suggestion to leverage the superpower nuclear reductions (i.e., six months before the end of a fixed duration, nations could give notice of their intent to withdraw from the treaty) was retained after a fashion in Article X. The six-month option was rejected along with Nigerian demands that the NPT explicitly empower members to withdraw if the treaty's disarmament aims were "being frustrated."[51] But it was agreed that the treaty would not be of indefinite duration. Instead, it would last 25 years and be reviewed as to whether or not it should be extended and, if so, how. As the Swiss noted, it was "preferable" that the treaty be "concluded for a definite period" so as to avoid "tying" the hands of nonweapons states who could not be expected to wait indefinitely on the weapons states to disarm.[52] Thus, any party to the treaty, under Article X, retained the right to withdraw if it "decides that extraordinary events, related to the subject matter of this treaty, have jeop-

---

51. See "Statement by the Nigerian Representative [Sule Kolo] to the Eighteen Nation Disarmament Committee: Draft Nonproliferation Treaty. November 2, 1967 [extract]," in ibid., pp. 557-58.

52. See "Swiss Aide Memoire to the Co-Chairmen of the Eighteen Nation Disarmament Committee: Draft Nonproliferation Treaty, November 17, 1967," in ibid., p. 573.

ardized the supreme interests of its country."[53]

Which Past as Prologue?

Reading the NPT today, one can easily forget that the original bargain of the Irish resolutions of the late 1950s and early 1960s is present in the final version of the NPT. Indeed, Articles I and II, which prohibit the direct or indirect transfer and receipt of nuclear weapons, nuclear explosives, or control over such devices, read very much like the original Irish resolutions themselves. In Article III, the treaty also calls on parties to accept and negotiate a system of safeguards that would prevent "Diversion of nuclear energy from peaceful uses to nuclear weapons or other nuclear explosive devices." Finally, the treaty makes it clear in Article IV that parties to the NPT could exercise their right to develop peaceful nuclear energy only "in conformity with Articles I and II."

Nor did the NPT's framers abandon their original concerns about the threat of catalytic or accidental nuclear war. The Germans in 1967, for example, defended the NPT aims "Because it is frightening to think what would happen if possession of nuclear weapons were spread chaotically through the world, if some adventurous state were one day irresponsibly to use such a weapon." Echoing this view, Germany's foreign minister argued that "even only one additional nuclear power would start a chain reaction that would be

---

53. On the point that this language meant that nonweapons nations might be compelled to withdraw if the weapons states did not live up to their pledge to disarm, see, for example, "Statement by the Swedish Representative [Alva Myrdal] to the Eighteen Nation Disarmament Committee: Nonproliferation of Nuclear Weapons, February 8, 1968;" "Statement by the Ethiopian Representative [Makonnen] to the First Committee of the General Assembly: Nonproliferation of Nuclear Weapons, May 6, 1968;" and "Statement by the Indian Representative [Husain] to the Eighteen Nation Disarmament Committee: Nonproliferation of Nuclear Weapons, February 27, 1968," in U.S. Arms Control and Disarmament Agency, *Documents on Disarmament, 1968*, Washington, DC: Government Printing Office, 1969, pp. 45, 293-94, and 116.

hard to control."[54] The Canadians made essentially the same point, arguing that some discrimination against nonweapons states was "the only alternative to allowing the continued spread of nuclear weapons...and such a process in the end would have no other result than nuclear war ... on the greatest scale."[55] The British representative to the General Assembly was just as emphatic:

> We are concerned not only that new possessors of nuclear weapons may employ them against each other, or against a non-nuclear state; we see an even greater danger in the possibility that the use of nuclear weapons by a third country could precipitate a war which would end in a nuclear exchange between the two so-called Superpowers. In our view, and I would think in that of the Soviet Union as well, each additional nuclear power increases the possibility of nuclear war, by design, by miscalculation, or even by accident.[56]

Competing against these concerns, however, was the view expressed by the Indian delegation

> that further proliferation is only the consequence of past and present proliferation and that unless we halt the actual and current proliferation of nuclear weapons, it will not be possible to deal effectively with the problematic danger of further proliferation among

---

54. See "Television Interview with Chancellor Kiesinger: Nonproliferation Negotiations, February 17, 1967 [extract)" and "Statement by Foreign Minister Brandt to the Bundestag on Proposed Nonproliferation Treaty, April 27, 1967," in *Documents on Disarmament, 1967*, pp. 91 and 215.

55. See "Statement by the Canadian Representative [Bums] to the Eighteen Nation Disarmament Committee: Nonproliferation of Nuclear Weapons, August 3, 1967," in ibid., p. 315.

56. "Statement by the British Representative [Hope] to the First Committee of the General Assembly, December 14, 1967," in ibid., p. 458.

additional countries.[57]

This alternative view, along with the idea that non-nuclear nations had inalienable rights to develop civilian nuclear energy and to withdraw from the NPT (and thus acquire nuclear weapons legally) if the superpowers did not disarm (or their security interests were at serious risk), became the NPT consensus view and was captured in Articles IV, V, VI, and X as well as most of the NPT's preamble.

Articles I and II, in contrast, reflected the original bargain of the Irish resolutions, which were concerned about the threat of accidental and catalytic nuclear war, whereas the NPT's other articles (with the possible exception of Article III) generally reflected the finite deterrence theorizing of the time.

The problem is that these two views are at odds. Certainly, it's difficult to argue that the further spread of even small numbers of nuclear weapons to other nations will significantly increase the risk of accidental or catalytic nuclear war, and at the same time recommend that nonweapons states threaten to acquire such weapons to get weapons states to limit their own nuclear arsenals. Yet, this is precisely the tension present in the negotiations leading up to the NPT and is reflected in the treaty's text (i.e., Articles I, II, VI, and X).

More important, this tension continues to be reflected in the debate over what constitute "peaceful" nuclear development in conformity with Articles I and II under Article IV. Nations who subscribed to the notion that the superpower arms race was a key cause of horizontal proliferation believed that nonweapons states deserved access to any and all civilian nuclear energy transfers to compensate them for their restraint and to assure them equal access to technology that the states with nuclear weapons already had.

---

57. See "Statement by the Indian Representative [Trivedi] to the Eighteen Nation Disarmament Committee: Nonproliferation of Nuclear Weapons, September 28, 1967," in ibid., p. 432.

For most of these nations, any civilian nuclear transfer made under safeguards was automatically "in conformity with Articles I and II." Indeed, for the Dutch, Belgians, and Luxembourgians—and, at times, even the Americans—the line between safeguarded and unsafeguarded activities under the NPT was, as one nonproliferation expert noted, "quite bright."[58] In May of 1968, the representative of the Netherlands government, for example, urged the superpowers to live up to their disarmament obligations under Article VI and explained that the obligation of nonweapons states to forgo the acquisition of nuclear weapons should "in no way" restrict their access to civil nuclear technology:

> My delegation interprets Article I of the draft treaty to mean that assistance by supplying knowledge, materials and equipment cannot be denied to a non-nuclear-weapon State until it is clearly established that such assistance will be used for the manufacture of nuclear weapons or other nuclear devices. In other words, in all cases where the recipient parties to the treaty have conformed with the provisions of Article III, there should be a clear presumption that the assistance rendered will not be used for the manufacture of nuclear weapons other explosive devices.[59]

The Americans were just as insistent that "Peaceful applications of energy derived from controlled and sustained nuclear reactions—that is, reactions stopping far short of explosion [had] nothing to do with nuclear weapons" and, thus, development of such applications

---

58. See, for example, Eldon V. C. Greenberg, *The NPT and Plutonium: Application of NPT Prohibitions to "Civilian" Nuclear Equipment, Technology and Materials Associated with Reprocessing and Plutonium Use*, Washington, DC: Nuclear Control Institute, 1993, pp. 18-19.

59. See "Statement by the Dutch Representative [Eschauzier] to the First Committee of the General Assembly: Nonproliferation of Nuclear Weapons, May 6, 1968 [extract]," in *Documents on Disarmament, 1968*, pp. 295-96.

would not be affected by the NPT's prohibitions.[60]

Yet, other evidence indicates that the NPT's framers felt uncomfortable about obligating the nuclear powers to provide any and all forms of nuclear-energy technology or materials, save nuclear explosives themselves. In the final debates over the NPT, Spanish and Mexican attempts to create a duty on the part of the nuclear haves to provide nuclear-energy aid to the nuclear have-nots and to reference "the entire technology of reactors and fuels" in the NPT's text were rejected.[61] This rejection, it has been argued, suggests that the NPT's framers understood that some forms of civil nuclear energy (e.g., weapons-usable nuclear fuels and their related production facilities) were so close to bomb making that sharing them might not be in "conformity" with Articles I and II.

More important, safeguarding such dangerous activities and materials was probably impossible. Certainly, inspections that lived up to Article III's requirement to "avoid hampering" nations' "technological development" and that remained in accordance with the NPT's concern—registered in its preamble—of focusing on the "flow" of source and special fissionable materials at "certain strategic points" would have difficulty accounting for significant quantities of weapons-usable materials at enrichment and reprocessing facilities, at reactors that used weapons-usable fuels, and at their respective fuel-fabrication plants. Nor would timely warning of diversions be likely. As ostensible "safeguards," such materials and activities would only mask the probable transfer or acquisition of nuclear weapons and thus violate the NPT's prohibitions in Articles

---

60. See "Statement by ACDA Director Foster to the First Committee of the General Assembly: Nonproliferation of Nuclear Weapons, November 9, 1966," in ibid., p. 721.

61. See "Spanish Memorandum to the Co-Chairmen of the Eighteen Nation Disarmament Committee, February 8, 1968," in ibid., p. 40 and "Mexican Working Paper Submitted to the Eighteen Nation Disarmament Committee: Suggested Additions to Draft Nonproliferation Treaty, September 19,1967," in *Documents on Disarmament, 1967*, pp. 394-95.

I and II and Article III's stricture that safeguards serve the purpose of verifying member nations' fulfillment of their NPT obligations.[62]

It would be nice if the NPT's negotiating record could settle such disputes. Unfortunately, it only raises them. Indeed, tension between the first three articles and those that follow in the NPT still exists today. Unaligned nations such as Indonesia and Mexico still argue that weapons states must go much further in reducing their nuclear arsenals and in sharing the benefits of peaceful nuclear energy to keep nonweapons states from abandoning the NPT. And the issue of just what constitutes effective safeguards under the treaty for trouble nations such as Iran and for dangerous nuclear activities such as reprocessing in Japan is as much a concern as ever.

A number of things, however, have changed since 1968. Instead of a superpower rivalry, only one superpower remains—the United States. Rather than an ever-escalating nuclear arms race, the United States and former Soviet republics are cooperating in reducing the number of nuclear weapons.

As for the promised benefits of peaceful nuclear power, these too seem less compelling. Certainly, few—if any—nations now believe that PNEs promise any economic benefits. The United States, India, and Russia—the only nations to experiment with such devices—no longer use them, and even Brazil and Argentina, who initially rejected the NPT because it would not allow them to acquire such devices, have renounced their development. Economically viable nuclear electricity, meanwhile, has been limited to uranium-fueled thermal reactors operating only in the most advanced economies of North America, Europe, and East Asia. The economical use of weapons-usable plutonium or mixed-oxide fuels in thermal or fast reactors is,

---

62. For this interpretation, see Greenberg; and Arthur Steiner, "Article IV and the 'Straightforward Bargain,'" PAN Paper 78-832-08, in Wohlstetter et al., *Towards a New Consensus on Nuclear Technology*, Vol. 2, Supporting Papers, ACDA Report no. PH-78-04-832-33, Marina del Rey, CA: Pan Heuristics, 1977.

at best, still many decades away.[63]

Meanwhile, the security dangers of certain types of civilian nuclear power and of reactor development in certain regions have become all too apparent. Iran, Saudi Arabia, Turkey, Algeria, and Egypt all have nuclear energy programs that are monitored by the IAEA. Yet, all harbor a desire to develop nuclear weapons and have attempted to evade IAEA inspections and proper import procedures. It is unclear if even special IAEA inspections could provide sufficient warning of dangerous activities in these politically turbulent nations.[64] IAEA monitoring of plutonium fabrication and reprocessing activities in such stable nations as Japan has also been criticized as being dangerously deficient. In fact, the amount of weapons-usable materials produced by such plants threatens to exceed the amount of fissile material present in the arsenals of weapon states.[65]

Finally, there is a newfound awareness that finite deterrence and the supposed stability that might come from threatening to attack an opponent's cities are nowhere near as sound as once supposed—either in theory or practice. The release of new information on the Cold War suggests that nuclear deterrence even between the superpowers was anything but automatic or guaranteed. Indeed, a nuclear incident in Cuba and/or possible war over intermediate-range nuclear force (INF) deployments in Europe was far more likely than many people imagined.[66]

---

63. For the latest economic forecast as to when such fuels might make economic sense, see, for example, Brian G. Chow and Kenneth A. Solomon, *Limiting the Spread of Weapon-Usable Fissile Materials*, Santa Monica, Calif.: RAND, October 1993, pp. 25-54.

64. See David Kay, "Detection and Denial: Iraq and Beyond," Washington, DC: Consortia for the Study of Intelligence, June 1994.

65. See, for example, Chow and Solomon, pp. xiv-xv; and Paul Leventhal, "IAEA's Safeguards Shortcomings—A Critique," Washington, DC: Nuclear Control Institute, September 12, 1994.

66. See, for example, William T. Lee, "The Nuclear Brink That Wasn't—and the One That Was," *The Washington Times*, February 7, 1995, p. A19.

Nor has finite deterrence proved to be as cheap or easy as originally promised. In the case of the French—the original innovators of finite deterrence—developing and maintaining a *force de frappe* has required spending billions of dollars annually to field several generations of strategic forces that have never seemed quite credible (or survivable enough) to other members of NATO—even against a limited Soviet attack. Smaller nations aiming to deter their near-nuclear neighbors or existing weapons states are likely to face similar challenges that proportionally will be at least as stressful.

These developments, of course, do not change the NPT's negotiating history. But they do suggest the relative risks of emphasizing NPT framers' concerns of the late 1960s over those they originally had in 1958. More important, by focusing on the NPT's original concerns, we are more likely to correct for its current deficiencies, which are themselves rooted in views that were all too popular at the time of its signing. Indeed, how well we focus on these concerns today will determine what worth the NPT will have in the decade ahead.

Chapter 2

How We've Come to View the NPT: Three Pillars

Dean Rust

For more than 45 years, international cooperation to prevent the spread of nuclear weapons has centered on the Nuclear Nonproliferation Treaty (NPT). The NPT was opened for signature on July 1, 1968, and entered into force on March 5, 1970. Momentum for the treaty stemmed from fears that the risk of nuclear war would increase as more countries acquired nuclear weapons. By the mid-1960s, five countries had conducted nuclear explosive tests and had incorporated nuclear weapons into their military forces.

The goal of the NPT was to "draw a line in the sand" at five nuclear countries. The treaty created two groups of states: Nuclear weapon states were those that had exploded a nuclear explosive device prior to January 1, 1967, (The United States, Soviet Union (USSR), United Kingdom (UK), France, and China); all others were non-nuclear weapon states. Non-nuclear states that join the treaty are prohibited from the manufacture or acquisition of nuclear weapons (Article II) and required to accept international verification of that undertaking (Article III). The weapon states that join are barred from assisting any non-weapon state to acquire nuclear weapons and from transferring weapons to others (Article I). In addition to these core nonproliferation obligations, the non-nuclear powers insisted the final version of the treaty preserve the right of all parties to a peaceful nuclear program (Article IV.1), facilitate peaceful nuclear cooperation among the parties (Article IV.2), and obligate all parties to pursue good faith negotiations on effective measures relating to nuclear disarmament and general and complete disarmament (Article VI). Throughout most of its history, the NPT has been defined by these three sets of obligations, which in shorthand are referred to as: Nonproliferation, peaceful nuclear uses, and disarmament.

Despite its imperfections and the tumultuous events of the past 45 years, the NPT has flourished and its membership stands at 189 parties (as of March 2016).[1] The treaty codified and has largely sustained the 1960s consensus against the spread of nuclear weapons. The norm it established led to the formation of other international tools which, along with the NPT, constitute the totality of the nuclear nonproliferation regime. Only four additional countries are credited with nuclear weapons over this period—India, Pakistan, North Korea, and Israel (although Israel has never declared the possession of nuclear weapons). This rate of proliferation is far less than many predicted in the 1960s, and the regime and standards fostered by the NPT deserve a major share of the credit. However, events of the last 25 years have led to unprecedented challenges to the treaty, including violations of its nonproliferation undertakings, the possible global expansion of civil nuclear energy, and post-Cold War expectations for credible progress toward nuclear disarmament. These difficulties are not entirely unexpected for this now 45 year old treaty, but how they play out in the coming years will have a major impact on the NPT's effectiveness and sustainability.

In 2010, the eighth conference of NPT parties (held every five years since 1970) was convened to review treaty implementation. While the parties expressed concerns about a range of issues, the treaty received a vote of confidence and an action plan to improve its operation was approved. The administration's approach to the NPT was laid out by U.S. President Barack Obama in an April 5, 2009,

---

1. This number does not include North Korea and Palestine, which are counted on some other lists.

speech in Prague: *"The basic bargain is sound: Countries with nuclear weapons will move towards disarmament, countries without nuclear weapons will not acquire them, and all countries can access peaceful nuclear energy."*[2] *In subsequent documents and speeches, the administration embraced the term "three pillars" to describe these elements of the "bargain" and portrayed them as "mutually reinforcing and interrelated." Moreover, disarmament was frequently cited first.* The aggregate of these factors (in italics for emphasis) constitutes the "three pillars strategy/approach" that is the subject of this chapter.

Describing the NPT in this manner has increased in recent years and chairs of NPT meetings have embraced it in their summaries as early as 2004. However, this approach by the Obama Administration represented a departure for the United States. Never before had the United States accorded disarmament this degree of emphasis at a Review Conference nor appeared to elevate its status equal to that of nonproliferation in terms of the treaty's viability. Given the positive outcome of the 2010 Conference, the administration can argue that the three pillars approach worked and that it was appropriate given the challenges to the treaty. But this approach also raises a number of questions.

Referring to the NPT as comprising pillars of nonproliferation, disarmament, and peaceful use conjures the image of three vertical structures of equal strength holding up the NPT, thus implying that weakness or failure of any one of the pillars would cause the treaty to topple. While such a literal meaning need not be presumed, the unqualified U.S. acceptance of the NPT as a bargain between the nuclear and non-nuclear powers and of the pillars as mutually reinforcing is consistent with such an image.

So why was this approach adopted? Does it alter the legal interpretation of the NPT? If not a different legal interpretation, then what

---

2. "Remarks by President Barack Obama," Hradcany Square, Prague, Czech Republic, April 5, 2009, available from *https://www.gpo.gov/fdsys/pkg/DCPD-200900228/pdf/DCPD-200900228.pdf.*

are the consequences of granting disarmament and peaceful uses a status equivalent to nonproliferation under the NPT? Are we bowing to the wishes of NPT nonaligned states that have long chafed at the preeminence given to nonproliferation? Is the degree of interdependence among the pillars suggested by this approach consistent with the treaty's structure or its history? Is it good or bad for the treaty in the long run? Or is it no big deal?

To frame the issue, this chapter offers a brief look at the NPT's negotiating history, its entry into force, and implementation up to the 2010 Conference. It shows how support developed in the 1960s for a treaty to prevent the spread of nuclear weapons to additional countries, and how disarmament and peaceful uses came to be included in the final version. It highlights the first 25 years of implementation (1970-1995) as a period of steady growth in NPT membership and acceptance of its nonproliferation obligations, but of scant progress on disarmament. Yet, despite this lag in Article VI implementation, the parties overwhelmingly consented to indefinite extension of the treaty in 1995—a clear endorsement of nonproliferation as the treaty's central purpose as intended by its founders.

The chapter then looks at NPT implement from 1995 to 2015, noting that pressure grew sharply for progress on Article VI—buttressed by arguments that the nonproliferation goals of the NPT had been largely fulfilled and that it was time for the nuclear powers to meet their side of the NPT "bargain," particularly given the end of the Cold War. At the same time, violations of the NPT's nonproliferation obligations were emerging. Serious differences developed among NPT parties over the relative importance of these disarmament and nonproliferation challenges, exacerbating tensions among the parties and contributing to the most dysfunctional review conference in the NPT's history in 2005. The Obama administration's embrace of the three pillars strategy can be seen as an attempt to restore a degree of consensus behind NPT implementation.

This history is then followed by an analysis. The author concludes that the three pillars strategy is not a reinterpretation of the NPT and that it was not an unreasonable choice for the Obama administration given what it perceived as the stakes in 2010. However, there are risks to this notion of the treaty; only time will tell whether they are serious and/or manageable. The NPT was conceived as a nonproliferation security pact among its parties, not as a means to achieve nuclear disarmament or to guarantee peaceful uses. The obligations in the treaty for each of the three undertakings are different—deliberately so. Perhaps the relative success of the NPT's nonproliferation mission has shifted expectations, but the United States must be careful that the three pillars strategy does not lead parties to demand more from the treaty on disarmament and peaceful uses than it can deliver. The pillars are not as interrelated as commonly assumed, and insisting on equal and balanced progress on each could weaken support for the treaty as a whole.

To mitigate these risks, the United States should ensure that its future NPT diplomacy fosters a greater appreciation of nonproliferation as the treaty's central purpose. There is ample evidence that proliferation abets insecurity (e.g. North Korea and Iran) and that the NPT is foremost a global barrier to contain these dangers short of war. While efforts with North Korea have failed to date, the NPT proved essential in finding a negotiated settlement to Iran's violations. At the same time, the United States should not relax its commitments under the treaty to peaceful uses and nuclear disarmament. Without credible U.S. policies to advance these objectives, support for the NPT nonproliferation undertakings could atrophy over time.

Negotiating History/Entry into Force

This section focuses briefly on the history, negotiation, and U.S. ratification of the NPT in relation to nonproliferation, disarmament, and peaceful uses.

1945 to 1965

The devastating consequences of World War II, including the destruction of Hiroshima and Nagasaki, defined early efforts "to save succeeding generations from the scourges of war, which in our lifetime has brought untold sorrow to mankind."[3] The first resolution adopted by the United Nations (UN) General Assembly on January 24, 1946, established an Atomic Energy Commission to make proposals "for control of atomic energy …to ensure its use only for peaceful purposes" and "for the elimination from national armaments of atomic weapons…"[4] The U.S. Acheson-Lilienthal and Baruch plans, which became the basis of negotiation at the UN Commission during 1946-1948, sought to ban nuclear weapons and to preserve for all nations the use of atomic energy for peaceful purposes under international control.[5] Thus, the early reaction to the invention of nuclear weapons was to outlaw them; they were viewed as incompatible with global security as that goal was defined in the aftermath of World War II. The Cold War dealt a predictable death to the Baruch Plan and its failure has resonated through the decades.

Preventing arms races of all kinds dominated the international disarmament agenda in the 1950s and up through 1962 as East-West

---

3. *Charter of the United Nations*, first preambular paragraph, entered into force October 24, 1945, available from *http://www.un.org/en/sections/un-charter/preamble/index.html*.

4. United Nations General Assembly, Resolution 1 (1), "Establishment of a Commission to Deal with the Problems Raised by the Discovery of Atomic Energy," January 24, 1946, available from *www.un.org/en/ga/search/view_doc.asp?symbol=A/RES/1(I)*.

5. Among the multitude of sources describing these ill-fated negotiations are Henry D. Sokolski, *Best of Intentions: America's Campaign Against Strategic Weapons Proliferation*, Westport, CT: Praeger, 2001, Chapter 2; Joseph Cirincione, *Bomb Scare: The History and Future of Nuclear Weapons*, New York: Columbia University Press, 2007, pp. 14-20; William Walker, *A Perpetual Menace: Nuclear Weapons and International Order*, New York: Routledge, 2012, pp. 44-48; and George Bunn, *Arms Control by Committee: Managing Negotiations with the Russians*, Stanford, CA: Stanford University Press, 1992, pp. 59-61.

tensions led to ever increasing stockpiles of weapons. General and complete disarmament proposals were discussed at the UN and in groups created for that purpose (e.g. Eighteen Nation Disarmament Committee located in Geneva). These schemes, similar to proposals following World War I, called for a phased and verified approach to reductions of all weapons' types.[6] States without nuclear weapons would pledge not to acquire them and states with nuclear weapons would gradually eliminate them. The disarmament proposals were generally well-conceived and allowed the United States and Soviet Union to nurture an image of sincerity, while the underlying tensions of the period meant the prospects for agreement were nil. Meanwhile, both countries were conducting aggressive programs to test and produce nuclear weapons.

U.S. President Dwight Eisenhower sought to channel the quest for peaceful uses of nuclear energy into a cooperative network to share these technologies in exchange for acceptance of international safeguards to guard against their use for nuclear explosives. This so-called "Atoms for Peace" proposal led to the establishment of the International Atomic Energy Agency (IAEA) in 1957 and to a flurry of U.S. exports of research reactors around the world. But IAEA safeguards applied only to these transfers; Atoms for Peace did not prevent countries from having indigenous facilities outside of safeguards or from pursuing a parallel nuclear weapons program.

By the end of the 1950s, it was clear that a comprehensive treaty that banned nuclear weapons proliferation and established a timetable for nuclear disarmament (by states then having nuclear weapons, i.e. the United States, USSR, and UK) was unachievable. Discussion of a separate nonproliferation agreement began to increase.

---

6. U.S. Arms Control and Disarmament Agency, *1962 Annual Report to Congress*, pp. 8-17, 54-83; also see *Documents on Disarmament 1946-1959* for UN resolutions calling for a phased program of disarmament, U.S. Department of State, *Documents on Disarmament 1945-1959*, Volume I, Department of State Publication 7008, Washington, DC: U.S. Government Printing Office, August 1960, pp. 383-384, 645-646, available from *https://www.un.org/disarmament/publications/documents-on-disarmament/1945-1956-dod/*.

There was virtual consensus that further nuclear proliferation would irreparably harm the security of all nations by increasing the risk of nuclear war, whether by accident or otherwise. Regional security would be undermined if competitors or adversaries sought to acquire nuclear weapons. Finally, it was clear that the goal of nuclear disarmament would become even more distant if the number of nuclear weapon states grew.

From 1958-1961 Ireland took the lead at the United Nations in focusing attention on these risks. By the end of 1961 Ireland had achieved unanimous support for a resolution in which the UN General Assembly "believing in the necessity of an international agreement… calls upon all states, and in particular upon the States at present possessing nuclear weapons, to use their best endeavors to secure the conclusion of an international [nonproliferation] agreement."[7] The essence of the nonproliferation agreement would require (i) weapon states not to transfer control of nuclear weapons to others and (ii) non-weapon states not to acquire or manufacture nuclear weapons.

Despite the growing consensus for a nonproliferation agreement, there was a foreshadowing in 1961 that gaining support from non-nuclear powers to abstain from acquiring nuclear weapons would not be unconditional. A Swedish resolution passed in December of that year called "for an inquiry into the conditions under which countries not possessing nuclear weapons might be willing to enter into specific undertakings to refrain from manufacturing or otherwise acquiring such weapons…" The results of this inquiry suggested that among these conditions would be commitments from the nuclear powers.[8]

---

7. Bunn, pp. 64-66; "The Treaty on the Nonproliferation of Nuclear Weapons," in Henry Sokolski, ed., *Reviewing the Nuclear Nonproliferation Treaty*, Carlisle, PA: Strategic Studies Institute, 2010, pp. 17-26, and Henry Sokolski, "What Does the History of the Nuclear Nonproliferation Treaty Tell Us About its Future?" in *Reviewing the Nuclear Nonproliferation Treaty*, pp. 31-38, available from *www.npolicy.org/thebook.php?bid=2*.

8. UN Department of Political and Security Council Affairs, *The United Nations*

Little of substance was accomplished over the next three years due to key Soviet concerns. Firstly, there were differences over whether a nonproliferation agreement should prohibit the stationing of nuclear weapons on the territories of non-nuclear powers—even if such weapons remained under the control of the nuclear power. Championed by the USSR, this ban on stationing was strongly opposed by the United States as undermining the North Atlantic Treaty Organization's (NATO) collective defense arrangements. Subsequently, these differences focused on the idea, then being discussed, of creating a NATO multilateral nuclear force, which the USSR viewed as incompatible with the emerging nonproliferation agreement. Secondly, the Soviets pressed to ensure that the non-weapon state members of the European Atomic Energy Authority (then comprised of Belgium, the Netherlands, Luxembourg, West Germany, Italy and which had its own safeguards system), would be subject to independent verification by the IAEA. These two roadblocks were ultimately removed: the idea of a multilateral nuclear force died on its own by 1965-66 and a compromise on safeguards was reached that allowed the IAEA to draw its own independent conclusions about compliance by the European non-nuclear powers with their NPT obligations.

The End Game–Addressing Peaceful Uses and Nuclear Disarmament

Meanwhile, by 1965 more countries at the UN and the Geneva disarmament conference, including those in the non-aligned movement, became actively engaged in the concept of a nonproliferation agreement. Once the idea took hold on the international disarmament agenda, it drew strong support among all nations as a top priority. Even representatives of nonaligned states were making

---

and Disarmament 1945-1970, New York: United Nations, June 1970, pp. 263-266, available from *https://www.un.org/disarmament/publications/yearbook/volume-1945-1970/*; also see discussion of the Swedish resolution in Sokolski, *Best of Intentions*, pp. 45, 47.

statements, filing joint memoranda and supporting UN resolutions declaring that the proliferation of nuclear weapons would endanger the security of all nations and urging the early conclusion of a treaty to prevent such proliferation.[9] The United States and Soviet Union each tabled the draft of a nonproliferation treaty in the summer of 1965.

This momentum culminated in the passage on November 19, 1965, of UN General Assembly resolution 2028 entitled "Non-proliferation of Nuclear Weapons" by a vote of 93-0 with five abstentions. It was sponsored by eight nonaligned nations and urged "all States to take all steps necessary for the early conclusion of a treaty to prevent the proliferation of nuclear weapons." It also made clear that the treaty "should be a step toward the achievement of …nuclear disarmament" and "should embody an acceptable balance of mutual responsibilities and obligations of both nuclear and non-nuclear powers."[10]

In 1966, it became increasingly clear that any nonproliferation treaty would have to address nuclear disarmament in some manner. As has been well-documented elsewhere, concerns about proliferation that had motivated the 1958-1961 discussion about a nonproliferation treaty were matched from 1965 onward by concerns about the U.S.-USSR nuclear arms race, which had shown no signs of abating de-

---

9. "Statement by the U.A.R. Representative (Hassan) to the Eighteen Nation Disarmament Committee, August 17, 1965," and "Eight Nation Joint Memorandum Submitted to the Eighteen Nation Disarmament Committee: Comprehensive Test Ban Treaty, September 15, 1965" in U.S. Arms Control and Disarmament Agency, *Documents on Disarmament – 1965*, ACDA Publication 34, Washington, DC: U.S. Government Printing Office, December 1966, pp. 345-346 and 424-426, both available from *http://unoda-web.s3-accelerate.amazonaws.com/wp-content/ uploads/assets/publications/documents_on_disarmament/1965/DoD_1965.pdf.*

10. "Statement by ACDA Director Foster to the First Committee of the General Assembly: World Disarmament Conference, November 18, 1965," in *Documents on Disarmament – 1965*, pp. 532-534.

spite the 1963 treaty to ban nuclear testing in the atmosphere.[11] In March 1966, the United Arab Republic echoed the views of many by suggesting that the treaty contain a separate article in which the nuclear powers would accept a legal obligation to halt the nuclear arms race and reduce stockpiles.[12] Later that year, eight nonaligned members of the Eighteen Nation Disarmament Committee outlined their views in a joint memorandum, stating that steps toward nuclear disarmament could be embodied in the treaty or as a declaration of intention. Chief among these steps were a comprehensive nuclear test ban, a treaty to halt the production of fissile material for nuclear weapons, and limits on and reductions of existing weapon stockpiles and means of delivery.[13] The United States and Soviet Union opposed the inclusion of specific steps, but suggested that general language related to nuclear disarmament could be included in the treaty's preamble.

The peaceful use issue had drawn little attention in the negotiations through 1966 for the obvious reason that the 1965 UN negotiating mandate for the NPT said nothing about peaceful uses. In 1967, however, West Germany, Mexico, the United Arab Republic, and Brazil began to stress that the treaty's prohibition on the acquisition of nuclear weapons should not impede the right of non-nuclear powers to realize the benefits of peaceful nuclear programs.[14] West

---

11. Sokolski, *Best of Intentions*, pp. 39-56; Sokolski, "What Does the History of the Nuclear Nonproliferation Treaty Tell Us About its Future?"

12. "Statement by the U.A.R. Representative (Khallaf) to the Eighteen Nation Disarmament Committee: Nonproliferation of Nuclear Weapons, March 3, 1966," in U.S. Arms Control and Disarmament Agency, *Documents on Disarmament-1966*, ACDA Publication 46, Washington, DC: U.S. Government Printing Office, September 1967, pp. 74-77, available from *http://unoda-web.s3-accelerate.amazonaws.com/wp-content/uploads/assets/publications/documents_on_disarmament/1966/DoD_1966.pdf*.

13. "Eight Nation Joint Memorandum Submitted to the Eighteen Nation Disarmament Committee: Nonproliferation of Nuclear Weapons, August 19, 1966," in *Documents on Disarmament-1966,* pp. 576-579.

14. See documents from: the Federal Republic of Germany on pp. 49, 51, 92,

Germany was particularly active in these discussions, including at the highest levels. Mexico made an impassioned plea for a treaty provision that would call for assistance to developing countries in this area.

The United States got the message; in a February 21, 1967, message President Johnson instructed U.S. negotiators "to exercise the greatest care that the treaty not hinder the non-nuclear powers in their development of nuclear energy for peaceful purposes"[15] and the identical U.S. and Soviet drafts tabled on August 24, 1967, included a new Article IV that looked very similar to the final version.

> Nothing in this treaty shall be interpreted as affecting the inalienable right of all the Parties to the Treaty to develop research, production and use of nuclear energy for peaceful purposes without discrimination in conformity with Article I and II of this Treaty, as well as the right of the Parties to participate in the fullest possible exchange of information for, and to contribute alone or in cooperation with other States to, the further development of the applications of nuclear energy for peaceful purposes.[16]

In presenting the text, U.S. Arms Control Agency Director Foster noted that Article IV was in response to several suggestions from non-nuclear powers and pointed out that the idea was drawn in part from Article 17 of the Latin American Nuclear Weapons Free Zone

---

95, 181, 211-212; Brazil on pp. 138, 226-227; the United Arab Republic, pp. 159, 424; and Mexico on pp. 394-398 in U.S. Arms Control and Disarmament Agency, *Documents on Disarmament – 1967*, ACDA Publication 46, Washington, DC: U.S. Government Printing Office, July 1968, available from *https://www.un.org/disarmament/publications/documents-on-disarmament/1967-dod/*.

15. "Message From President Johnson to the Eighteen Nation Disarmament Committee, February 21, 1967, in *Documents on Disarmament – 1967*, p. 98.

16. "Draft Treaty on the Nonproliferation of Nuclear Weapons, August 24, 1967," in *Documents on Disarmament – 1967*, p. 340.

Treaty that had been opened for signature in February of 1967.[17] Article 17 of that treaty reads:

> Nothing in the provisions of this Treaty shall prejudice the rights of the Contracting Parties, in conformity with this Treaty, to use nuclear energy for peaceful purposes, in particular for their economic development and social progress.[18]

In contrast to peaceful uses, language related to nuclear disarmament in the August 24, 1967, identical U.S. and USSR drafts remained in the preamble. While improving on such language in earlier drafts, it was still quite modest:

> Declaring their intention to achieve at the earliest possible date the cessation of the nuclear arms race,
>
> Urging the cooperation of all States in the attainment of this objective,
>
> Desiring to further the easing of international tension and the strengthening of trust between States in order to facilitate the cessation of the manufacture of nuclear weapons, the liquidation of all their existing stockpiles, and the elimination from national arsenals of nuclear weapons and their means of delivery pursuant to a treaty on general and complete disarmament under strict and effective international control[19]

---

17. "Statement by ACDA Director Foster to the Eighteen Nation Disarmament Committee: Draft Nonproliferation Treaty, August 24, 1967," *Documents on Disarmament – 1967*, pp. 344-345.

18. "Treaty for the Prohibition of Nuclear Weapons in Latin America, February 14, 1967," in *Documents on Disarmament – 1967*, p.77.

19. "Draft Treaty on the Nonproliferation of Nuclear Weapons, August 24, 1967," in Documents on Disarmament – 1967, p. 339.

While this language gave clear priority to halting the nuclear arms race, it essentially consigned the goal of eliminating nuclear weapons to the realm of science fiction by stating that it would be achieved *"pursuant to a treaty on general and complete disarmament"* (emphasis added).

This formulation was criticized because it was not in the main body of the treaty and because it was tied to general and complete disarmament. On September 19, 1967, Mexico acknowledged that the treaty could not obligate the nuclear powers "to conclude future disarmament agreements" (general or specific); to do so "would be tantamount to opposing the very existence of a non-proliferation treaty." However, the statement continued, the nuclear powers should be willing in the treaty to "pursue negotiations in good faith in order to conclude such agreements"[20] and to do so without linkage to general and complete disarmament. Mexico offered the following language for the operative part of the treaty:

> Each nuclear-weapon State Party to this Treaty undertakes to pursue negotiations in good faith, with all speed and perseverance, to arrive at further agreements regarding the prohibition of nuclear weapon tests, the cessation of the manufacture of nuclear weapons, the liquidation of their existing stockpiles, the elimination from nation arsenals of nuclear weapons and the means of their delivery, as well as to reach agreement on a Treaty on General and Complete Disarmament under strict and effective international control.[21]

---

20. "Statement by the Mexican Representative (Castaneda) to the Eighteen Nation Disarmament Committee: Nonproliferation of Nuclear Weapons, September 19, 1967," in *Documents on Disarmament – 1967*, p. 400.

21. "Mexican Working Paper Submitted to the Eighteen Nation Disarmament Committee: Suggested Additions to Draft Nonproliferation Treaty, September 19,1967," *Documents on Disarmament – 1967*, p. 395.

On January 18, 1968, a new joint U.S.-Soviet draft of the treaty was tabled and it contained a paragraph on disarmament in the operative section that borrowed ideas from the Mexican draft.

> Each of the Parties to this Treaty undertakes to pursue negotiations in good faith on effective measures regarding cessation of the nuclear arms race and disarmament, and on a treaty on general and complete disarmament under strict and effective international control.[22]

Importantly, it made clear that the obligation to pursue these goals fell on all parties to the treaty, not just the nuclear powers. The draft omitted the specific measures from the Mexican draft. U.S. and Soviet negotiators agreed that an obligation to achieve any particular agreement was a step too far. However, in explaining the new text, Arms Control Agency Deputy Director Adrian Fisher conceded to an interpretation favored by many non-weapon states. First, he accepted that the purpose of the negotiations called for in the new operative paragraph is the *conclusion* (emphasis added) of disarmament agreements; and, second, he acknowledged that the draft article does not require the negotiation of nuclear arms control measures to occur within the framework of a treaty on general and complete disarmament. To quote Fisher:

> As Mr. Castaneda [of Mexico] has pointed out, although the nuclear Powers cannot actually undertake to conclude particular future disarmament agreements among themselves at this stage, they can undertake to initiate and pursue negotiations in good faith in order to conclude such agreements. That is essentially the content which has been given

---

22. "Revised Draft Treaty on the Nonproliferation of Nuclear Weapons, January 18, 1968," in U.S. Arms Control and Disarmament Agency, *Documents on Disarmament – 1968*, ACDA Publication 52, Washington: DC, U.S. Government Printing Office, September 1969, p. 4, available from *https://www.un.org/disarmament/publications/documents-on-disarmament/1968-dod/*.

to the obligation which we are recommending be incorporated into the body of the Treaty. The new draft article constitutes a solemn affirmation of the responsibility of nuclear-weapon State to strive for effective measures regarding cessation of the nuclear arms race and disarmament. Moreover, the article does not make the negotiation of these measures conditional upon their inclusion within the framework of a treaty on general and complete disarmament.[23]

Before leaving Article VI, it is important to understand the context of how it came to be included in the treaty. Since the 1950s the nuclear weapon states had agreed in principle on the need for measures to control nuclear weapons. While these efforts did not succeed in the context of broader disarmament schemes, after the 1962 Cuban Missile Crisis the United States and Soviet Union began to focus on separate issues where agreement seemed possible and to begin a step-by-step approach. In 1963, the hot line agreement and limited test ban treaty were concluded. And through the rest of the decade the two sides, with the United States in the lead, increased their efforts to find ways to limit both offensive and defensive strategic forces.

This U.S.-Soviet discussion was happening at the same time as momentum built for the NPT. So a general undertaking by the nuclear powers along the lines of Article VI was compatible with what the United States and Soviet Union were already doing. The point is that the United States did not view Article VI as creating a new obligation or as something it would undertake as a "favor" for the non-nuclear powers. As the first nuclear weapon state, the United States had long ago accepted the responsibility to find ways to avoid nuclear war and to pursue negotiations to that end.

This view was stressed by a U.S. official in a statement to the Ge-

---

23. "Statement by ACDA Deputy Director Fisher to the Eighteen Nation Disarmament Committee: Draft Nonproliferation Treaty, January 18, 1968," *Documents on Disarmament–1968*, p. 15.

neva conference on February 6, 1968. It is quoted below at length to reinforce the point, but also as a rejoinder to the idea that the United States viewed Article VI as a quid-pro-quo for non-weapon states abjuring nuclear weapons.

> The United States shares the prevailing view of the importance of provisions in this treaty dealing with a cessation of the nuclear arms race. However, this is not a question of making some compensating sacrifice; we believe it is in our national interests to halt the nuclear arms race and to begin reducing existing nuclear arsenals...
>
> The tendency to view a commitment to nuclear disarmament by the nuclear weapon states as a quid pro quo for the renunciation of nuclear weapons by other states fails to take into account the actual intention and situation of the overwhelming majority of non-nuclear-weapon states...the vast majority of such states have no intention, desire or indeed any early prospect of producing or acquiring nuclear weapons or other nuclear explosive devices. Moreover, those who look for a quid pro quo seem to consider this treaty as if it were a commercial contract in which each party seeks to trade off concessions in order to gain equal financial or trade benefits. However, the nonproliferation treaty is not that kind of agreement; its primary benefit accrues to all of us directly in the form of enhanced security and not as a result of balanced concessions....
>
> It seems quite evident that the primary benefit conferred by this treaty is the assurance it provides, in the first instance, to the non-nuclear-weapon states that their non-nuclear neighbors or rivals will not have to assume the enormous expenditures, and the

serious security risk, of acquiring nuclear weapons.[24]

Both Articles IV and VI were tweaked slightly over the next four months; notably the word "nuclear" was inserted before "disarmament" in Article VI's first clause. A preambular paragraph was added that recalled past expressions of support for a treaty to ban the testing of nuclear weapons.

On Article IV, it is important to note that a Nigerian official complained on May 8, 1968, that the final version of Article IV.2 created no obligation on NPT parties to provide "full scientific and technological information" to non-weapon states in peaceful uses. This admission demonstrates that nonaligned countries, despite frequent charges against nuclear supplier countries over the last 40 years of violating Article IV.2, were fully aware that it did not mandate unconditional peaceful nuclear assistance. The relevant quote from the Nigerian official:

> nuclear-weapon Powers do not accept any obligation to furnish non-nuclear States with full scientific and technological information in their possession on the peaceful uses of nuclear energy...this article does not deal adequately with the problem of bridging the intellectual gap which would be created by the restrictions imposed by Articles I and II on development of nuclear energy for peaceful purposes.[25]

The final version of the NPT was presented to the General Assembly on May 31, 1968. General Assembly resolution 2373, approved on June 12, 1968, by a vote of 95-4-21 (abstentions), commended the NPT and expressed "the hope for the widest possible adherence....

---

24. "Statement by the United States Representative (De Palma) to the Eighteen Nation Disarmament Committee: Nonproliferation of Nuclear Weapons, February 6, 1968," in *Documents on Disarmament–1968*, pp. 36-38.

25. "Statement by the Nigerian Representative (Ogbu) to the First Committee of the General Assembly: Nonproliferation of Nuclear Weapons, May 8, 1968," in *Documents on Disarmament–1968*, p. 300.

by both nuclear-weapon and non-nuclear-weapon states."[26]

Review, Extension, and Withdrawal

Before leaving the negotiating phase, it is useful to highlight other provisions of the NPT that were viewed as a means to ensure weapon state accountability under Article VI. The requirement to hold a conference after five years (Article VIII.3) to review implementation of the treaty was clearly viewed in part as a means to take stock of efforts toward ending the nuclear arms race and achieving nuclear disarmament.

Italy proposed during negotiations that the nonproliferation treaty be of definite duration with periodic renewals, while allowing any party the right not to renew with six months advance notification (the U.S.-Soviet draft had called for an NPT of unlimited duration).[27] This approach would have offered non-nuclear powers an easy "out" should the nuclear powers not fulfill their Article VI obligations. This proposal was not accepted, although it was a factor in creating Article X.2 which set an initial 25 year duration for the NPT, after which a vote would be taken on whether to extend the treaty indefinitely or for a fixed period(s).[28]

There is some mention in the negotiating history of efforts to make withdrawal "easier," but nothing substantive came of them.[29] Swe-

---

26. "General Assembly Resolution 2373 (XXII): Treaty on the Nonproliferation of Nuclear Weapons, June 12,1968," in *Documents on Disarmament–1968*, p. 431.

27. "Statement by the Italian Representative (Caracciolo) to the Eighteen Nation Disarmament Committee: Draft Nonproliferation Treaty, October 24, 1967," in *Documents on Disarmament-1967*, pp. 527-529.

28. George Bunn, "The NPT and Options for its Extension," *Nonproliferation Review*, Vol. 1, No. 2, Winter 1994, pp. 53-54.

29. "Statement by the Nigerian Representative (Sule Kolo) to the Eighteen Nation Disarmament Committee: Draft Nonproliferation Treaty (Extract), Novem-

den said in early 1968 that it seemed "reasonable" that if the treaty review process led to a conclusion that the nuclear powers had "blatantly disregarded" their Article VI undertakings, then non-nuclear parties might be justified in withdrawing.[30] However, this linkage did not appear to generate much support before the treaty was presented in its final form to the UN General Assembly in May 1968.

George Bunn and Roland Timerbaev—members of the U.S. and Soviet delegations that negotiated the NPT—have written about the negotiating history of the withdrawal provision (Article X.1) which reads as follows:

> Each Party shall in exercising its national sovereignty have the right to withdraw from the Treaty if it decides that extraordinary events, related to the subject matter of this Treaty, have jeopardized the supreme interests of its country. It shall give notice of such withdrawal to all other Parties to the Treaty and to the United Nations Security Council three months in advance. Such notice shall include a statement of the extraordinary events it regards as having jeopardized its supreme interests.

In brief, they argue that the negotiators attempted to set a high standard, requiring the party intending to withdraw to provide a statement to all other NPT parties and to the Security Council describing the "extraordinary events" that had "jeopardized [its] supreme interests." The drafters believed that events leading an NPT party to withdraw could be so grave as to constitute a threat to international

---

ber 2, 1967," and "Nigerian Working Paper Submitted to the Eighteen Nation Disarmament Committee: Additions and Amendments to the Draft Nonproliferation Treaty, November 2, 1967"in *Documents on Disarmament-1967*, on pp. 557-558; and "Brazilian Amendments to the Draft Treaty on Nonproliferation of Nuclear Weapons, February 13, 1968," on p. 65, in *Documents on Disarmament-1968*, pp. 557-558.

30. See "Statements by Swedish Representative Alva Myrdal on February 8 and March 5, 1968," in *Documents on Disarmament-1968*, pp. 45,151-152.

peace and security, including that of its neighbors, and thus within the jurisdiction of the Security Council.[31] Implied in this reasoning is that frustration with Article VI implementation would not by itself meet that standard. While some may question this conclusion today based on the history of Article VI implementation, from the negotiating history it is clear that the periodic five year review and 25 year duration (with extension options) were viewed as the principal means for holding the nuclear powers accountable.

U.S. Ratification and Entry into Force

On July 1, 1968, the treaty was opened for signature in the capitals of the three depositary governments—Washington, Moscow, London—and 66 countries signed on that day. As stated in Article IX.3, the NPT would enter into force once 43 signatory states had deposited an instrument of ratification, including the three depositary governments. As a prime supporter of the treaty, the United States moved quickly. The treaty was sent to the U.S. Senate for advice and consent to ratification on July 9, 1968, and the Foreign Relations Committee held hearings on July 10, 11, 12, and 17. The Committee reported the treaty favorably in September (13-3 with 3 abstentions), but the political crisis stemming from the Soviet invasion of Czechoslovakia in August 1968 deferred floor action for the rest of the year.

---

31. See George Bunn, Roland Timerbaev, James Leonard, "Nuclear Disarmament: How Much Have the Five Nuclear Powers Promised in the Non-Proliferation Treaty?" *Lawyers Alliance for World Security*, June 1994, pp.19-22, available from *cisac.fsi.stanford.edu/sites/default/files/Bunn_Nuclear_Disarmament.pdf*; George Bunn and Roland Timerbaev, "The Right to Withdraw from the Nuclear Non-Proliferation Treaty (NPT): The Views of Two NPT Negotiators," *Yaderny Kontrol Digest*, Vol. 10, No. 1 and 2, Winter/Spring 2005, pp. 22-25, available from *fsi.stanford.edu/sites/default/files/Bunn_Timerbaev.pdf*; and George Bunn and John Rhinelander, "The Right to Withdraw from the NPT: Article X is Not Unconditional," *Disarmament Diplomacy*, Issue 79, April/May 2005, pp. 2-3, available from *www.acronym.org.uk/dd/dd79/79gbjr.htm*.

In 1968, the Committee's formal consideration of the treaty focused very little on Article VI. One possible explanation is that the undertaking in Article VI "to negotiate in good faith on effective measures related to nuclear disarmament" was consistent with what the United States was planning to do in any event.

With regard to peaceful uses, most of the Senate's attention was directed to Article V, which called for international arrangements to share in the benefits of nuclear explosions for peaceful purposes. The idea of digging canals or stimulating underground gas reservoirs using nuclear explosives was topical in the 1960s, but interest soon faded and Article V became effectively inoperative following the 1996 test ban treaty that outlawed all nuclear explosions. The Senate's 1968 discussion of this issue is not currently relevant to the NPT peaceful uses debate. However, in regard to other peaceful uses, the hearing record addressed five issues that are of continuing interest.

First: Administration testimony, particularly on July 12, 1968, from Glenn Seaborg, the Chairman of the Atomic Energy Commission (AEC), provided a glimpse into how the NPT was viewed given the expectation of a substantial growth in civil nuclear power in the coming years.[32] Seaborg painted the NPT as critical toward providing assurance that the many power reactors expected to come on line would be used only for peaceful purposes. Without the NPT, the only way these reactors would become subject to international nonproliferation commitments would be if a country invited the IAEA to apply safeguards or if safeguards were required by the government supplying the reactor or fuel.

Second: Seaborg said the United States already had a strong record of sharing the peaceful uses of nuclear energy. In a subsequent submission for the record on September 5, 1968, the administration listed 33 bilateral U.S. civil nuclear agreements then in force (many

---

32. "Statement by A.E.C. Chairman Seaborg to the Senate Foreign Relations Committee: Nonproliferation Treaty, July 12, 1968," in *Documents on Disarmament-1968*, on p. 516.

dating from the mid-1950s), and two multilateral agreements (with the IAEA and European Atomic Energy Agency).

Third: In response to a question during testimony about what constitutes the "manufacture of a nuclear explosive device," which is prohibited in Article II, Arms Control Agency Director William Foster provided an extension of his remarks including the following: "Neither uranium enrichment nor the stockpiling of fissionable material in connection with a peaceful program would violate Article II as long as these activities were safeguarded under Article III." However, the Foster statement also noted that placing an activity under safeguards "would not, in and of itself, settle the question of whether that activity was in compliance with the Treaty."[33]

Clearly, the United States was not prepared to "green light" any activity undertaken pursuant to the rights affirmed in Article IV.1 of the treaty even if subject to IAEA safeguards. The hearing record states that this position had also been affirmed by the United States during the negotiation of the treaty.

Fourth: AEC Chairman Seaborg, in response to a series of questions from the Committee, made clear on September 12, 1968, that Article IV.2, which calls for the "fullest possible exchange" of equipment, material and information for the peaceful uses of nuclear energy, does not compel the United States

> to embark on any costly new programs or as obliging the U.S. to meet all requests and demands...nor will it remove the discretion we have in determining the nature of our cooperative relationships with other countries, on a case by case basis. The words "fullest possible exchange" in Article IV clearly imply that the Parties will be expected to cooperate

---

33. U.S. Congress, "Nonproliferation Treaty," hearings before the Committee on Foreign Relations, United States Senate, 90th Cong., 2nd Sess., July 10, 11,12, and 17, 1968, Washington, DC: U.S. Government Printing Office, 1968, p. 39.

only to the extent that they are able to do so...[34]

Finally, IAEA safeguards on peaceful nuclear programs were in their infancy in 1968 and the Committee made clear in its report that their effectiveness under the NPT remained to be determined.

> The reliability and thereby the credibility of international safeguards systems is still to be determined. No completely satisfactory answer was given the committee on the effectiveness of the safeguards systems envisioned under the treaty...The committee is fully aware of the potential problems in the safeguards field. But it is equally convinced that when the possible problems in reaching satisfactory safeguards agreements are carefully weighed against the potential for a worldwide mandatory safeguards system, the comparison argues strongly in favor of the present language of the treaty.[35]

Following the November 1968 election, the new administration under U.S. President Richard Nixon endorsed the treaty and urged Senate action. Hearings were held February 1969 and the previous administration's views on NPT matters were reaffirmed. There were no new substantive disclosures, although more attention was paid to Article VI. One Article VI highlight came from a question and answer for the record provided to the Senate by the new Arms Control Agency Director Gerard Smith: "Article VI of the NPT merely requires us to 'pursue negotiations in good faith.' It does not require us to achieve any disarmament agreement, since it is obviously impossible to predict the exact nature and results of any such

---

34. "Letter From AEC Chairman Seaborg to Senator Cooper on the Nonproliferation Treaty, September 11, 1968," in *Documents on Disarmament-1968*, p. 636.

35. "Report by the Senate Foreign Relations Committee on the Treaty on the Nonproliferation of Nuclear Weapons, September 26, 1968," in *Documents on Disarmament-1968*, p. 656.

negotiations."[36]

This assertion tracks with the letter of the treaty, but the previous administration (see above) had publicly stated during the negotiations that the purpose of Article VI would be to "conclude" agreements. These two lines of interpretation were not necessarily inconsistent and the different emphasis is not surprising given the different contexts.

The Committee reported favorably on the NPT by a vote of 14-0-1 (present).

The Senate gave its advice and consent to ratification on March 13, 1969, with a vote of 83-15. The Committee report acknowledged the U.S. commitment in the treaty "to move to negotiate the means of limiting, if not ending, the nuclear arms race" and declared that "decisions facing both countries [United States and USSR] in the area of strategic offensive and defensive missiles are of vital importance not only to the peace and security of the world but to the successful implementation of the Nonproliferation Treaty."[37]

There was uncertainty about the future of the NPT despite the strong support it elicited. While nearly 100 countries signed the treaty between July 1, 1968, and early 1970, many had indicated there would be delays in ratification. It was not until March 5, 1970, that 43 ratifications were obtained and the NPT entered into force. Key U.S. allies such as West Germany and Japan were not among this group, as the details of IAEA safeguards application in their countries were still being negotiated. Others were waiting to see how much support the NPT would gain over time before committing.

---

36. "Answers to Questions by Senator Thurmond to ACDA Director Smith on the Nonproliferation Treaty, February 28, 1969," in *Documents on Disarmament-1969*, p. 53, available from *https://www.un.org/disarmament/publications/documents-on-disarmament/1969-dod/*.

37. "Report by the Senate Foreign Relations Committee on the Nonproliferation Treaty, March 6, 1969," in *Documents on Disarmament-1969*, p. 95.

Comments

Since 1945 world leaders had focused on ways to prevent a recurrence of Hiroshima and Nagasaki. The initial approach dealt with elimination and nonproliferation. But by the late 1950s, it seemed clear that East-West tensions would continue to thwart elimination by those already in possession of nuclear weapons; and the wiser course was to focus on keeping the nuclear club from growing. There was little dispute that the core provisions of such a nonproliferation agreement would obligate non-weapon states not to acquire nuclear weapons and weapon states not to transfer weapons to others. To obtain broader support, particularly from nonaligned states, the final version included articles on peaceful uses and disarmament, although each undertaking was general in nature.

The literal meaning of Article VI does not require the conclusion of agreements, but there was an expectation that the pursuit of "negotiations in good faith" would lead to concrete agreements and progress toward the goals of ending the nuclear arms race and of nuclear disarmament. The United States did not discourage such an expectation. The nonaligned states were explicit about the types of measures that should be pursued, although the United States and Soviet Union were careful not to endorse the idea that Article VI required the conclusion of any particular agreement.

While the NPT established no specific measures or deadlines in regard to Article VI, it was clear that the 5-year treaty review process and the 25-year duration provision were aimed partly at holding the nuclear powers accountable to their undertakings in this Article. The record makes clear that the United States and UK were fully aware that their Article VI undertaking would be extremely important to the viability of the treaty. Yet, the undertaking did not seem onerous at the time as the United States was already committed to engagement with the Soviet Union on strategic nuclear issues and Article VI promised nothing more than "to pursue negotiations in good faith." Moreover, no one could predict in 1968 how long the NPT would survive, let alone how salient Article VI would grow over the years.

On peaceful uses (apart from nuclear explosions for peaceful purposes), the negotiations led to the inclusion of a so-called savings clause, i.e. to reassure non-weapon states that by giving up the right to acquire nuclear weapons they were not also forfeiting their right to peaceful nuclear energy (Article IV.1). Clearly, however, the exercise of this right had to conform with the nonproliferation undertakings in Articles I, II, and by extension Article III. In general, the negotiators did not intend to prohibit any legitimate peaceful nuclear activity as long as it was accounted for under an IAEA safeguards agreement. However, the tension between the NPT's nonproliferation goals and certain peaceful applications, such as the use of plutonium as reactor fuel, was already evident. Countries such as Germany wanted reassurance that the NPT would not hinder its ambitious civil nuclear energy plans, and the United States reassured them—up to a point. No clear line was drawn between prohibited and permitted activities, which should not be surprising given the inherent duality of nuclear energy.

To compound this ambiguity, the IAEA safeguards system was in its infancy and it was not clear how effective it would be in detecting secret nuclear activities or the diversion of nuclear material from a declared peaceful nuclear program. The nonproliferation mission of the NPT faced a very uncertain future. In addition to questions about safeguards, it was impossible to know whether technologies relevant to nuclear proliferation could be effectively controlled, or how quickly these technologies would diffuse to others, or how the inevitable expansion of peaceful nuclear uses to other countries would influence proliferation risks.

The undertaking "to facilitate" cooperation among countries in the peaceful uses of nuclear energy (Article IV.2) under IAEA safeguards represented nothing more than what had already become routine under Atoms for Peace—at least for the United States. It was placed in the treaty because developing countries wanted to ensure that the NPT would not result in new obstacles to this cooperation. In the 1960s, the potential benefits of the then-growing field of civil nuclear applications seemed limitless and the developing countries did not want to lose out. This attitude was consistent with the prevailing view in the post-colonial era where the growing number of independent states argued that advanced countries had an obligation to share technology across the board to aid in economic development. That said, the language of Article IV.2 was crafted to give advanced nuclear nations flexibility on the scope and conditions of such cooperation.

The clear purpose of the NPT upon its entry into force was to prevent further proliferation. The negotiating history and the treaty's provisions clearly reflect that reality. Most non-weapon states, despite the inherent inequality of such an agreement, were on board. They believed it was in their self-interest to prevent proliferation; a world of increasing numbers of nuclear weapon states was viewed as making all nations less secure. Yet, they thought it only fair that a nonproliferation treaty should also include provisions to require those already with nuclear weapons to negotiate limits on their nuclear arsenals and to ensure no curtailment of the right to benefit from the peaceful uses of nuclear energy.

Treaty Implementation

We turn to the history of the treaty's implementation to draw lessons about how the parties in practice have viewed operation of the nonproliferation, nuclear disarmament, and peaceful use undertakings. The narrative through 1995 will be brief and provide an overview of the decision that year on extending the treaty. For 1995 onward, more detail is offered to provide context leading to the 2010 Conference and the "three pillars" strategy.

The First Twenty-Five Years: 1970-1995

While entry into force of the NPT in 1970 was accomplished with strong support from the international community, no one could predict how effective the treaty would be in halting proliferation beyond the five countries in possession of nuclear weapon to that point.

India's 1974 detonation of a so-called peaceful nuclear explosive was the first major blow to NPT goals. India had opposed the NPT and with this nuclear detonation made clear that it would not allow a discriminatory treaty such as the NPT to deny it entry into the exclusive nuclear club. India's explosion also revealed a major problem with the conditions nuclear suppliers had attached to their exports—India had used assistance from Canada and the United States in the manufacture of its nuclear explosive under the guise that it was for "peaceful purposes." However, if India intended to stall the global nonproliferation movement, it failed. Support for the NPT continued to grow and the Nuclear Suppliers Group was formed in 1975 to address weaknesses in nuclear export controls that India had exploited. This suppliers' initiative also led to the first instance of multilateral cooperation in reducing the spread of enrichment and reprocessing plants, nuclear facilities with civilian uses but which also can produce high enriched uranium and plutonium usable in nuclear weapons. While formed in response to India's test, this supplier group soon found itself also battling a secret Pakistani nuclear procurement program set up to match India's acquisition of a nuclear explosive.

Iraq's nuclear procurement activities in the 1970s suggested that it would be the first to challenge the NPT from within the ranks of treaty parties. Israel did not wait for any "smoking gun," deciding in 1981 to destroy two Iraqi research reactors supplied by France before they became operational. Despite this military action, Iraq continued to pursue an underground nuclear weapons program, including enrichment technologies, which went undiscovered until the IAEA was given enhanced inspection authority after Iraq's defeat in the 1991 Gulf War. Multiple NPT violations were soon exposed and remedial actions were approved by the UN Security Council.

North Korea got into the act less than two years later when it refused an IAEA inspection to investigate its declaration of nuclear material. Rather than address the violation, North Korea provided notification of its intention to withdraw from the NPT—an action that North Korea later suspended in June 1993 following talks with the United States. In November 1994, the two sides concluded a phased agreement that froze reprocessing and the production of plutonium, but deferred actions to bring North Korea into NPT compliance until a later date.

As with India's 1974 nuclear explosive test, the discovery of Iraq's nuclear weapons program in 1991 prompted strong nonproliferation countermeasures. The IAEA launched a multi-year program in 1993 to improve its safeguards system. Previously, the IAEA had focused primarily on deterring diversion from declared facilities rather than detecting secret ones. In 1992, the Nuclear Suppliers Group concluded a major upgrade of its nuclear export guidelines, including the addition of controls on nuclear dual-use items of the type Iraq had secretly acquired for its nuclear weapons program. Controls on enrichment and reprocessing items were expanded.

Global attention to Iraq's proliferation programs seemed to galvanize other emerging nonproliferation trends. In 1992, China and France joined the NPT—reversing more than 20 years of shunning the treaty. In 1993, political leaders in Argentina and Brazil decided to end the ambiguous status of their nuclear programs by bringing into force a comprehensive IAEA safeguards agreement. They acceded to the NPT later in the decade. An altered political landscape led South Africa to join in 1991 and to reveal in 1993 that it had manufactured six nuclear weapons, which had been dismantled prior to joining the treaty—the first known case of a country eliminating its nuclear weapons prior to joining the NPT.

Turning to key nuclear disarmament events during the NPT's first 25 years, the first major step was in 1972 when the United States and Soviet Union concluded agreements limiting offensive and defensive systems. Despite intensive efforts and a threshold test ban treaty in 1974, follow on efforts to further limit strategic offensive nuclear forces stalled for nearly two decades. A treaty was signed in 1979, but never entered into force. U.S.-Soviet relations were disrupted by the 1979 Soviet invasion of Afghanistan and by frequent changes in USSR leadership in the early 1980s. As the U.S. delegation to the 1985 Review Conference admitted, good faith was being exercised but the results were "disappointing."[38] This bleak record of Article VI implementation was finally broken in 1987 with a treaty banning U.S. and Soviet intermediate range nuclear forces. The end of the Cold War also led to major reductions in overseas deployments of tactical nuclear weapons in the early 1990s, and to unilateral steps by Russia and the United States to halt nuclear explosive testing. Negotiations on a nuclear test ban treaty, on and off since the 1950s, began in earnest in 1994. Also in 1994, the first treaty calling for reductions of U.S.-Russian strategic nuclear arms entered into force.

An underappreciated landmark during this period was the actions by the former republics of the Soviet Union to return nuclear weapons deployed on their territories to Russia and to join the NPT as non-nuclear powers. This outcome was particularly significant in the case of Belarus, Kazakhstan, and Ukraine, which had long-range nuclear missiles of the former USSR within their borders after independence. The fact that the 1991 breakup of the Soviet Union did not lead to the creation of new nuclear weapon states was a momentous contribution to global security. It was due in no small part to the existence of the NPT, which two decades prior

---

38. "Statement by ACDA Director Adelman to the Third Review Conference of the Parties to the NPT, August 28, 1985," in *Documents on Disarmament-1985*, p. 555, available from *http://unoda-web.s3-accelerate.amazonaws.com/wp-content/uploads/assets/publications/documents_on_disarmament/1985/DoD_1985.pdf*.

had declared nonproliferation a global norm and over the years had demonstrated its value as a treaty-based framework to codify and verify non-nuclear undertakings.

Article IV attracted attention periodically up to 1995 due to the tightening of nuclear export controls in response to the 1974 Indian nuclear test and to the 1991 discovery of Iraq's violations. These actions by the Nuclear Suppliers Group caused an uptick of concern over the NPT's Article IV.2 undertaking to cooperate in the peaceful uses of nuclear energy. However, despite the adoption of progressively stronger export controls, the United States and other nuclear suppliers were able to show credible implementation of Article IV.2, clearly demonstrating that NPT parties in compliance experienced few problems in obtaining assistance, whether bilaterally or through the IAEA. Moreover, the claim that supplier controls were a violation of Article IV.2 never gained serious traction.

In sum: Leading to the 1995 extension decision, the NPT's nonproliferation mission had gained a solid vote of confidence. NPT membership grew from 46 in 1970 to 175 in 1995. Proliferation in India and violations by Iraq and North Korea had led to concerted action by NPT parties to improve systemic barriers to proliferation. Several long-standing outliers had joined between 1990 and 1995. The NPT had provided a political and legal framework that helped to prevent new nuclear powers arising from the breakup of the Soviet Union. Moreover, the evident lack of progress on Article VI through 1991 appeared not to dampen the treaty's appeal. States continued to join the NPT and to stay in. The unsatisfactory outcomes of the 1980 and 1990 review conferences—due largely to Article VI concerns—did not lead any party to withdraw from the treaty. And the end of the Cold War offered the prospect of progress on Article VI that had been largely thwarted by East-West tensions during the NPT's first two decades.

Turning to the 1995 month-long NPT Review and Extension Conference, by the third week a clear majority of parties had publicly declared for indefinite extension. However, a few others favored a series of 25 year extensions and were prepared to block consensus on indefinite extension and force a vote.[39] While the outcome was not in doubt (per the NPT, the decision required a majority of parties), Conference leaders wanted to avoid taking a vote that would record a sizable number of treaty parties against indefinite extension. The Conference President proposed a way to bypass a vote by gaining consensus on the proposition that there was a majority in favor of extension. The treaty was extended indefinitely on May 10, 1995.[40]

This outcome came at a price, albeit a small one given the stakes. The legal decision to extend the treaty was accompanied by adoption of three political decisions: (i) substantive principles and objectives were set as future goals for treaty implementation, including the completion of negotiations on a test ban treaty by 1996; (ii) the five year review process was strengthened, and (iii) the creation of a weapons of mass destruction free zone in the Middle East was endorsed.[41] Many viewed the extension of the NPT and these understandings as a package; and without them support for indefinite extension would have been weaker. Taken together, these decisions set up a process to enhance accountability for NPT implementation, including for Articles IV and VI.

The fact that the NPT became permanent, despite a questionable Article VI record by the nuclear powers over 20 of those 25 years, demonstrated that the vast majority of treaty parties understood and valued its central purpose of preventing the spread of nuclear weapons. Yet, it was also clear that wide spread support for in-

---

39. Harald Mueller, "A Cornerstone of World Order: Extending the NPT," *NATO Review*, Vol. 43, No. 5, September 1995, pp. 21-26, available from *www.nato.int/docu/review/1995/9505-5.htm*.

40. The Final Document of the 1995 Review and Extension Conference of the Parties to the Treaty on the Non-proliferation of Nuclear Weapons. NPT/CONF.1995/32 (Part I), p. 12, available from *https://www.nonproliferation.org/wp-content/uploads/2015/04/2010_fd_part_i.pdf*.

41. Ibid, pp. 8-14.

definite extension was predicated on the prospect that the end of the Cold War would lead to rapid progress toward the goals of Article VI, including conclusion of a test ban treaty.

How the NPT Fared Following Indefinite Extension and Up through 2015

Developments

The 1995-2015 period of NPT implementation began under a promising set of circumstances. The treaty had become virtually universal and the parties had agreed to an enhanced treaty review process. Despite the Iraqi and North Korean violations, no other serious compliance problems had surfaced and efforts were underway to strengthen IAEA safeguards. The end of the Cold War had created a promising environment for long-delayed progress on Article VI.

The situation began to unravel even before 2000. While a treaty to ban nuclear explosive testing was concluded in 1996, it was rejected by the U.S. Senate in 1999. The promise of U.S.-Russia cooperation on further nuclear reductions fell victim to domestic factors in both countries, driven in part by missile defense issues. The effort to begin negotiations on a fissile material cutoff treaty stalled in the Conference on Disarmament.

In a huge blow to the twin goals of nonproliferation and nuclear disarmament, India carried out nuclear explosive tests in May 1998 and announced its intention to incorporate nuclear weapons into its military forces. Pakistan followed suit three weeks later. It had been 34 years since the last declared nuclear weapon state—China in 1964, a hiatus in proliferation due to the NPT and the norm it established. These actions did not constitute NPT violations as neither country had ever joined the treaty. However, this action by India was they were a direct challenge to the nonproliferation standard established by the treaty and also to regional security as countries that had fought three wars now launched into nuclear arms competition. Meanwhile, the Iraqi and North Korean violations continued to fester without resolution. Iraq's continued defiance led to U.S.-UK bombing of suspected proliferation sites in December 1998, which provided a pretext for Iraq to end UN-mandated inspections.

The negative turn of events continued from 2001-2009 under U.S. President George W. Bush. 9/11 led to a cascade of fears about nuclear terrorism and the possibility that rogue states like Iraq, Iran, or North Korea might facilitate such acts. By 2003, concern over Iraq's proliferation programs led to a U.S. led invasion that toppled Iraqi President Saddam Hussein. In August 2002, North Korea was confronted with evidence of a secret enrichment program, which led to its withdrawal from the NPT in 2003. In 2006, North Korea conducted its first nuclear weapon test.

In 2002, a secret nuclear program in Iran dating from the 1980s was exposed; it included uranium enrichment and the construction of a reactor optimized to produce plutonium. A subsequent IAEA investigation found multiple violations by Iran of its NPT safeguards agreement and uncovered an extensive nuclear black market run by A.Q. Khan, the father of Pakistan's nuclear weapons program. Libya too was implicated as a recipient of Khan's nuclear largesse. In 2003, Libya abandoned its proliferation programs, and Iran ended aspects of its nuclear weapons program.[42] However, Iran continued to pursue enrichment and cited as justification its "inalienable right" to a peaceful nuclear program under Article IV.1 of the NPT. This action defied both the IAEA Board of Governors and a 2006 UN Security Council resolution calling for the suspension of enrichment. Syria's failure to account for what the IAEA believed was the construction of a secret reactor, which Israel destroyed in 2006, further compounded the challenge to the NPT in the Middle East.

---

42. National Intelligence Estimate – Iran: Nuclear Intentions and Capabilities, November 2007, available from *http://www.dni.gov/files/documents/Newsroom/Reports%20and%20Pubs/20071203_release.pdf*.

Nonproliferation developments have been mixed under President Obama. No new NPT parties surfaced as threats. However, despite increasing sanctions, North Korea rebuffed any effort at dialogue and responded with nuclear tests in 2009, 2013 and 2016, and recently claimed to have miniaturized a nuclear warhead for delivery by ballistic missile. From 2009-2013, Iran's stockpile of enriched uranium continued to grow despite sanctions by the UN Security Council. A military attack on Iran's nuclear program seemed a definite possibility. Many thought it was only a matter of time before Saudi Arabia and others in the Middle East sought to match Iran's growing nuclear weapons capability. Yet, Iranian elections in 2013 led to a surprise breakthrough as the new government led by President Hassan Rouhani proved willing to negotiate a comprehensive settlement (described below) to the decade long impasse stemming from its NPT violation.

Responses

All three U.S. administrations pursued aggressive policies designed to bolster the NPT's nonproliferation pillar. The Clinton administration continued policies to improve IAEA safeguards, including the negotiation of a model IAEA safeguards additional protocol that was approved by the IAEA Board of Governors in 1997. This Protocol was designed to improve safeguards in NPT parties by granting enhanced legal authority to the IAEA to detect and deter clandestine nuclear activities. The administration also rallied a strong international condemnation by NPT parties of India's and Pakistan's 1998 acquisition of nuclear weapons, including the imposition of a wide range of sanctions (which turned out to be short-lived).

The Bush administration pushed strongly to punish violators and to establish new tools to thwart proliferation threats. The September 11, 2001, attacks also led to efforts to clamp down on the ability of terrorists to acquire material and equipment for nuclear weapons. Among these steps were UN-mandated rules for controlling proliferation-related items and enhanced cooperation with Russia to protect nuclear material from theft. These multiple efforts had the benefit of imposing another layer of protection against state-level proliferation.

In 2005, the United States joined the European Union (EU) in negotiations to resolve Iran's noncompliance. This effort was unsuccessful and led to three rounds of UN Security Council-mandated sanctions on Iran during the Bush administration. The United States refuted Iran's claim of a right to an unhindered peaceful nuclear program by pointing out that Article IV.1 stipulates that any such program must be compatible with the NPT's nonproliferation objectives. Aspects of Iran's program were cited as evidence to the contrary, and the United States concluded that "Article IV does not provide States Parties that have violated the nonproliferation provisions of the treaty any protection from the consequences of breach."[43] This conclusion was accepted by key NPT parties and served to undergird UN sanctions on Iran.

Six-party talks with North Korea aimed at restoring its compliance with the NPT seemed promising at times, but any agreements soon foundered in implementation. North Korea's withdrawal, the first from the treaty, also prompted a major discussion among NPT parties of its legality and how to deter and respond to any future withdrawal actions by others.

There was a setback on Article VI as the Bush Administration decided to pursue unilateral actions on U.S. nuclear forces and not be tied down by treaties. While these policies still led to substantial U.S. reductions and to reduced reliance on nuclear deterrence, the reaction of NPT parties was largely negative. This approach rejected many of the Article VI measures that had been adopted by consensus at the 1995 and 2000 review conferences with the sup-

---

43. Christopher Ford, "NPT Article IV: Peaceful Uses of Nuclear Energy" Statement to the 2005 Review Conference of the Treaty on the Nonproliferation of Nuclear Weapons, New York, May 18, 2005, available from *https://2001-2009.state.gov/t/vci/rls/rm/46604.htm*.

port of the Clinton Administration. In a further slap at the NPT, the United States decided in 2005 to exempt non-NPT party India from a nearly 30 year U.S. ban on sales of nuclear fuel to that country—a policy devised in the 1970's to provide significant civil nuclear assistance only to NPT parties in recognition of their willingness to forswear nuclear weapons. Indeed, this NPT preference policy had been embraced by nearly all nuclear fuel suppliers by 1992 and continually recognized by NPT fuel recipients as an important benefit of Treaty membership.

The Obama administration placed preventing proliferation and nuclear terrorism at the top of America's nuclear agenda. It continued many of the nonproliferation policies pursued by the previous administration, while tweaking some of them. In particular, efforts to obtain a legally-binding commitment to forego enrichment and reprocessing from states not already in possession of them were modified on a case-by-case basis. The President stressed the importance of consequences for those who break the rules. To that end, the administration achieved a substantial ratcheting up of sanctions against Iran in 2010 and steadily increasing pressure on North Korea.

While North Korea showed no evidence of a willingness to negotiate, the Iranian government elected in 2013 proved willing to limit its nuclear program in exchange for sanctions relief. The final deal among Iran and the United States, UK, Russia, China, France, Germany and the European Union was concluded in July 2015 and began implementation in January 2016. It calls for intrusive verification along with unprecedented limits on Iran's nuclear program lasting 15-25 years. To the extent this arrangement proves effective in avoiding or at least delaying further proliferation in the Middle East, it has to be considered a very positive achievement for the NPT.

In nuclear arms control, President Obama departed from the previous administration by declaring U.S. support for "the peace and security of a world without nuclear weapons," signing a New Strategic

Arms Reduction Treaty (START) treaty in 2010, and expanding U.S. policy against the use of nuclear weapons. The administration also embraced the "grand bargain" concept of the NPT and described nonproliferation, nuclear disarmament, and peaceful uses as "mutually reinforcing pillars of the NPT." Moreover, in an unprecedented gesture to non-nuclear states' concerns about Article VI, the President's policy statements on the NPT placed nuclear disarmament ahead of nonproliferation when discussing the three "pillars."[44] However, lack of support in the U.S. Congress for steps beyond the new START Treaty and a deterioration in U.S.-Russian relations effectively thwarted other aspects of the President's nuclear agenda as of this writing (March 2016). There was no progress with Russia on further reductions nor in the Senate on U.S. ratification of the 1996 nuclear test ban treaty and of non-use Protocols to three regional nuclear weapon free zone treaties (for the South Pacific, Africa and Central Asia) submitted during the Obama administration.

Review Conferences

The outcomes of the four Review Conferences over this period reflect in large measure the policy views of the Clinton, Bush, and

---

44. See Barack Obama, "Remarks by President Barack Obama," Hradcany Square, Prague, Czech Republic, April 5, 2009, available from *http://iipdigital.usembassy.gov/st/english/texttrans/2009/04/20090406115740eaifas0.9701763.html#axzz4XYKTJZrX*; Rose Gottemoeller, "Opening Statement at the Third Session of the Preparatory Committee for the 2010 Nuclear Non-Proliferation Treaty Review Conference," UN Headquarters, New York, May 5, 2009, available from *http://iipdigital.usembassy.gov/st/english/texttrans/2009/05/2009050 8114236eaifas0.9583704.html#axzz4XYKTJZrX*; Barack Obama, "Statement by Barack Obama on the Release of the Nuclear Posture Review," April 6, 2010, available from *https://www.gpo.gov/fdsys/pkg/DCPD-201000235/pdf/DCPD-201000235.pdf*; and State Department Fact Sheet, "The Nuclear Non-Proliferation Treaty, Promoting Disarmament," April 27, 2010, available from *hhttp://iipdigital.usembassy.gov/st/english/article/2010/04/20100430172028xjsnomm is0.1379772.html#axzz4XYKTJZrX*.

Obama administrations toward NPT implementation. Due to the Clinton administration's flexibility on Article VI, the NPT received a strong endorsement by its parties at the 2000 NPT Review Conference in what was generally seen as a balanced review of nonproliferation, nuclear disarmament, and peaceful uses. Included in that 2000 outcome was a set of expectations about future implementation, which the parties believed would advance the purposes of the NPT. Among these measures were "13 steps" related to Article VI.[45]

The run up to the 2005 Conference, as noted above, included a number of negative developments for the treaty in all three components. Not surprisingly, the 2005 Conference was the most contentious and least productive on record. Iran's presence guaranteed substantial controversy over Article IV and that any effort to single out its noncompliance would be blocked. U.S. efforts to focus on nonproliferation were met with pushback by non-nuclear powers who believed the U.S. had reneged on important Article VI commitments.

Going into the 2010 Conference, President Obama's reversal of his predecessor's approach to nuclear arms control prompted renewed hope and a spirit of compromise. While the Conference was unable to agree on the review of the last five years, it did reach consensus on a forward-looking action plan. That said, the action plan was nuclear disarmament-heavy and nonproliferation-light. This outcome may have been inevitable given Iran's inclination to veto any new nonproliferation measures and the desire of many NPT parties to make up for what was seen as an erosion in U.S. support for Article VI implementation during the previous administration.

The prospects for the April/May 2015 Conference were dim as President Obama's 2009 Prague agenda had proved too ambitious as noted above. The Iran nuclear issue was still unresolved at the time and North Korea's bellicosity had further dampened any hopes for

---

45. Final Document from the 2000 Review Conference of the Parties to the Treaty on the Non-Proliferation of Nuclear Weapons, NPT/CONF.2000/28, Part I, pp. 14-15, available from *www.reachingcriticalwill.org/images/documents/Disarmament-fora/npt/GENERAL-DOCS/2000FD.pdf*.

a roll-back of its nuclear program. Nonetheless, substantial agreement was reached on many key issues. However, in the end, the United States and a few other states could not join consensus on a final document due to the failure to find compromise language on steps to convene a conference related to establishing a WMD free zone in the Middle East.

Comments

From 1995-2015, the NPT's nonproliferation mission experienced its first sustained test of enforcement. The Iraqi and North Korean cases began before 1995, but work to resolve them continued well into this period—with efforts to contain North Korea since 2006 focused on denuclearization. Adding the Iranian, Libyan and Syrian violations created an unprecedented challenge to the treaty's effectiveness in containing proliferation. The vast majority of NPT parties proved willing to confront noncompliance by non-nuclear powers and to utilize the IAEA and, if necessary, the UN Security Council to address violations. Violations by Iraq, North Korea, Iran, Libya, and Syria were all, to varying degrees, dealt with by these international institutions. The problems in enforcement stemmed from differences among NPT parties about the pace, scope and effectiveness of actions to resolve the violations. This task was compounded by the failure of the violators, with the exception of Libya, to cooperate fully. The difficulties of forcing sovereign states headed by autocratic leaders to reveal their nuclear secrets became painfully evident.

As noted earlier, these frustrations led twice to the use of force, both times in Iraq. A four-day bombing campaign of suspected proliferation sites was carried out by the United States and United Kingdom in December 1998 and a U.S.-UK led invasion to overthrow Saddam Hussein was undertaken in March 2003. These actions were conducted without explicit authorization of the UN Security Council, although the second was far more consequential for the NPT.

The 2003 invasion reflected a decision to abandon inspections and sanctions as a means to enforce compliance in favor of forcible regime change. This action was widely criticized by other countries, including many U.S. allies. Other NPT parties saw the United States as unwilling to trust internationally-sanctioned measures to enforce the treaty; U.S. leadership on the NPT suffered accordingly. This negative view of the United States deepened as no evidence was found after the invasion of a reconstituted nuclear weapons program, as the United States had alleged.

Such a use of military force to resolve proliferation violations without the authorization of the UN Security Council will inevitably complicate future efforts to gain international support for diplomacy aimed at resolving compliance matters. Why would any violator caught cheating be willing to pursue a negotiated solution when the United States has a past record of rejecting these outcomes in favor of military action? Similarly, potential supporters of a diplomatic solution might be reluctant to sign up if the United States is considered unreliable in terms of its willingness to search for and support a peaceful resolution of an NPT violation. That does not obviate the right of the United States to act unilaterally in self-defense, but to do so in the absence of an imminent threat would inevitably impose serious costs on U.S. global leadership in this field.

Indeed, Iran's violation and failure to promptly resolve all questions led to frequent public discussion about the use of military force—with Israeli Prime Minister Benjamin Netanyahu barely concealing his desire for the United States to launch such an attack. Such a step was resisted by both the Bush and Obama administrations, despite Iran's continued failure to engage in constructive negotiations to address the demands of the UN Security Council.

Whether the peaceful resolution of Iran's NPT violation proves successful in the long run, only time will tell. It came at a high price—too high for some—in that Iran was allowed to retain an enrichment program—albeit substantially limited for 15 years. On the other hand, it set a helpful precedent that NPT noncompliance can lead to enforcement action against a state's nuclear program, even those peaceful activities claimed to be a sovereign right under NPT Article IV. Iran had fought mightily against such action, but consistently lost this argument at both the IAEA and UN Security Council. At a minimum, the deal has delayed what appeared in 2012-2013 to be a steady march toward the acquisition of enough enriched uranium for Iran to acquire a nuclear weapon within four to six weeks of a decision to do so. The deal lengthened that time period to one year, adequate time to rally substantial opposition to a resumption of Iran's nuclear weapons ambition.

It demonstrates that with resolve and the right political conditions, the NPT can be effectively enforced without resort to military action. The Treaty's essential role in fostering global security received a large boost—without the NPT and its IAEA safeguards system, there would have been no foundation for the kind of global pressure through the UN Security Council that was applied to Iran. Despite opponents' claims, the outcome represented a strong endorsement of the NPT's nonproliferation mission. Any deal like this has risks, but so would military action. Whether the potential downsides of the deal will occur and/or be effectively managed will not be known for many years. The history of the nonproliferation regime shows many instances of imperfect solutions that nonetheless buy time and delay the onset of nuclear risks.

Events of the past 25 years have revealed deficiencies in NPT enforcement. In the first instance, increasing the capacity of the IAEA "watch dog" to bark early and often remains a work in progress. Moreover, while the IAEA has been persistent once unleashed, it is too methodical for the United States in pursuing violations. Prompt and satisfactory NPT enforcement will likely occur only in cases where the state in question comes clean and fully cooperates with Libya being the prime example.

Should there be future cases like Iraq and Iran, however, it will take long and sustained action and even then the resolution may not be fully satisfactory. But if the outcome at least substantially delays the time by which a country could acquire nuclear weapons, it can still be viewed positively as having created space for an eventual strategic decision by the country in question to remain compliant with the NPT and forego nuclear weapons. Moreover, as Iraq demonstrates, target state resistance does not necessarily mean that the IAEA has failed, nor should it translate into a "worst case" scenario that presumes a continued march to nuclear weapons. The downside, of course, is that long delays in NPT enforcement foster uncertainty about the violator's true intentions. This situation in turn could stimulate the very actions that peaceful enforcement actions are designed to forestall, such as the use of military force against the suspect facilities or a regional competitor seeking its own nuclear weapon capabilities. With Iran, these negative consequences did not occur despite its decade long refusal to resolve the violations. Finally, with time running out and military action increasingly likely, sanctions began to impose ever greater costs and dramatic changes in Iranian political leadership created an opening for a peaceful resolution. It is hoped that Iran will comply with the deal and that the downsides of the delay in its resolution will continue to be avoided.

While serious violations by five NPT parties tested the treaty, these challenges were met with strong action, if not always textbook in their implementation. In the one case where force was used, Iraq, it turned out to be unnecessary. Libya cooperated and continued international pressure on Iran appears to have halted its march to nuclear weapons, at least for 15-20 years assuming it complies with the basic undertaking of the 2015 joint plan. Syria remains on the docket, but further action to resolve its violation seems highly unlikely except in the context of resolving the ongoing civil war. Meanwhile, it seems equally unlikely that Assad would take the risk of committing further NPT violations with the future of his country and his leadership already in jeopardy.

North Korea withdrew from the NPT and is building and testing nuclear weapons. It represents the one clear failure of enforcement and poses a major danger to that region of the world. To date, its situation and behavior point to this as a "one off" action and not indicative of a systemic problem with the treaty. It will inevitably lead to talk of South Korea and/or Japan starting their own nuclear weapon programs, actions that would be a severe blow to the NPT. So far, however, the United States appears to have credibly reinforced its extended deterrent relationship so that neither government is considering abandoning the NPT.

Other trends since 1995 demonstrate continued support for the NPT's nonproliferation component. Importantly, virtually all parties remain in compliance and no new countries have been found in violation for nearly a decade. Cuba finally joined in 2002, deciding that its political differences with the United States over the Treaty were less important than the stigma of being an outlier to a virtually universal treaty. More than 120 NPT parties have brought into force the Additional Safeguards Protocol since 1997. The reluctance of a few holdouts like Egypt and Brazil to sign this Protocol obscures the fact that most NPT non-nuclear powers with significant nuclear activities have accepted it. This is an important gain for nonproliferation as the Protocol represents a major upgrade in NPT verification from the standard that applied during the treaty's first three decades.

The Nuclear Suppliers Group has continued to improve its export guidelines to respond to the ever-changing threat picture, and currently boasts 48 members, including several members of the non-aligned group. This too reinforces the NPT's nonproliferation mission as countries frequently critical of the suppliers group have now embraced its Guidelines. The fact is no state is immune from penetration by clandestine nuclear procurement networks, whether state-affiliated like A.Q. Khan or those sponsored by terrorist groups. Nuclear export controls remain a vital part of an NPT party's commitment to nonproliferation. Finally, the extensive cooperation since 9/11 to strengthen measures against the threat of nuclear terrorism has also contributed positively to nonproliferation.

The Middle East (ME) resolution adopted at the 1995 Conference should be noted briefly in this discussion of NPT party practices related to nonproliferation. The resolution urges the creation of a proliferation-free zone in the Middle East. It is widely viewed as the "price" for having gained Arab acquiescence in the consensus decision to extend the NPT indefinitely. The Arab League, under Egyptian leadership, uses the resolution to highlight Israel's failure to join the NPT and forswear nuclear weapons. At the 2010 Review Conference, the parties called for a separate Conference to be convened in 2012 to address the ME resolution. Despite the intensive efforts of a Finnish facilitator, the Conference has not been held and the issue proved to be a major negative factor at the 2015 review conference. This resolution contributes to the discussion of the NPT's nonproliferation objectives, but to date not in a constructive fashion. It has effectively institutionalized a high level of attention to Israel's non-party status while no comparable effort is made toward India and Pakistan.

Turning to nuclear disarmament, the heightened interest in Article VI since 1995 stems in part from the greater accountability imparted to the review process. Other factors include the Treaty's near universality, which has led many non-nuclear powers to argue that the NPT has largely fulfilled its promise of nonproliferation. They note that despite violations by a few countries more than 180 non-nuclear states remain in good standing. This narrative then turns to Article VI, and the degree to which its promise has been realized.

This emphasis on Article VI was buoyed in 1996 when the International Court of Justice, at the request of the UN General Assembly, issued an advisory opinion on the legality of the threat or use of nuclear weapons. The opinion included a finding that Article VI encompasses "an obligation to pursue in good faith and bring to a conclusion negotiations leading to nuclear disarmament."[46] As an advisory opinion, it does not have the force of law. The United States had unsuccessfully urged the Court to decline the request given that

---

46. International Court of Justice, "Legality of the Threat or Use of Nuclear Weapons," Advisory Opinion of July 8, 1996, available from *http://www.icj-cij.org/docket/files/95/7495.pdf*.

it was being asked to render an opinion on a hypothetical situation without any concrete set of facts.[47]

This growing emphasis on Article VI is also bolstered by the end of the Cold War and an anticipation of faster progress toward nuclear disarmament. The United States and Russia continue to feel most of the heat since they possess over 90% of the world-wide total of nuclear weapons. Unfortunately, while agreements and unilateral actions have substantially reduced stockpiles, serious political differences between and within the United States and Russia have obstructed the type of irreversible and verifiable treaties that non-nuclear powers believe are the hallmark of progress on Article VI. Since 2010, non-governmental organizations and nonaligned countries have focused on the humanitarian consequences of the use of nuclear weapons as a means of energizing progress toward nuclear disarmament. In general, non-governmental groups have seized on widespread sentiment that there are still far too many nuclear weapons and that their modernization is both unnecessary and too costly.

Article IV has gained prominence since 1995 largely because Iran used it as a defense against pressure to curtail its enrichment program. Its skillful diplomats have exploited the "inalienable right" language of the NPT to create sympathy among nonaligned countries for its position. Support for Iran's "victimization" narrative had been inadvertently stoked by international discussion of alternatives to the establishment of new national programs in enrichment and reprocessing. The United States and other NPT leaders rebutted Iran's arguments effectively while noting that NPT parties in good standing continued to receive Article IV benefits. While attention to this issue will continue at a moderate level, Article IV concerns do not trigger the depth and breadth of grievance that most non-nuclear states experience about Article VI.

---

47. Michael J. Matheson, "The Opinions of the International Court of Justice on the Threat or Use of Nuclear Weapons," *American Journal of International Law*, Vol. 91, July 1997, pp. 417-435.

In summary, events of the last 20 years have substantially increased the political importance and visibility across the board of all three treaty components—nonproliferation, disarmament, and peaceful uses. While nonproliferation retains its salience as the NPT's number one priority, it is clear that implementation of Article VI will demand greater attention. The role of Article IV in the NPT's future will depend to some extent on how states respond to the Iran settlement and the growth in civil nuclear uses worldwide. Any significant uptick in demand for the establishment of new national enrichment and reprocessing programs will conflict with current international nuclear export standards that strongly discourage such exports, and could lead to broader dissatisfaction with Article IV.

The Three Pillars: Analysis

The Policy

The previous section described how nuclear disarmament and peaceful uses have increased in prominence since 1995, and did so while nonproliferation also gained more attention due to serious cases of noncompliance. This context along with the NPT policies of the Clinton, Bush, and Obama administrations, and the outcomes of the three Review Conferences prior to 2010 offer a plausible explanation for the three pillars strategy.

At the 1995 and 2000 Review Conferences, the United States emphasized nonproliferation (per Iraqi and North Korean violations) while acknowledging that more should be done to move toward nuclear disarmament. In that context, the United States agreed to include in the Final Documents a list of steps that would represent progress on Article VI. These outcomes were consistent, as explained earlier, with the shift in focus after the Cold War toward Article VI implementation and with the Clinton administration's nuclear arms control agenda.

By the 2005 Conference, further instances of noncompliance had taken place (Libya, Iran, along with North Korean withdrawal) and the Bush Administration "doubled-down" on nonproliferation. This strategy was understandable in light of these events and arguably represented the strongest U.S. affirmation of the treaty's nonproliferation mission at a review conference in 20 years. However, as described above, it coincided with the Bush administration's turnabout from the nuclear arms control approach of the Clinton Administration and from the Article VI steps endorsed at the 1995 and 2000 Conferences.

Before looking at the U.S. approach to the 2010 Conference, recall that at Prague in 2009, President Obama said America had a moral responsibility to lead in reducing the dangers of nuclear weapons and outlined an ambitious agenda (described earlier). When it came time to translate these policies into a strategy for the 2010 NPT Review Conference, the Administration (a) embraced what it called the basic bargain of the NPT: Weapon states move to disarmament, non-weapon states forsake nuclear weapons; all parties abiding by their obligations gain access to peaceful uses; (b) adopted the term "pillars," which was already in common use by many NPT parties, to describe the treaty's three main components and gave priority to nuclear disarmament; and (c) accepted the three pillars as interrelated and mutually reinforcing. As noted at the outset, it is the totality of these factors that comprises the three pillars strategy for the purposes of this chapter.

A search of NPT records over its history would likely find "pillar" used on occasion to describe components of the NPT. But its use increased dramatically in the run up to the 2005 Conference. References can be found, for example, in statements from that period by the Non-Aligned Movement (NAM), Canada, and in the Chair's summary of the 2004 meeting of the Preparatory Committee. This summary states that the pillars "represent a set of interrelated and mutually reinforcing obligations and rights of States parties."[48]

---

48. Chairman's Summary, Preparatory Committee for the 2005 Review Confer-

Moving to the next review cycle, the Chair's working papers from meetings in 2007 and 2008 mention the NPT as resting on three pillars and that the nonproliferation and disarmament pillars are reinforcing. A new point in these 2007/2008 documents stresses "the importance of balanced, full and non-selective application and implementation of the Treaty." In contrast to the Chair's listing of pillars in 2004—which cited nonproliferation, disarmament, and peaceful nuclear cooperation, in that order—the 2007 and 2008 papers list nuclear disarmament, nonproliferation, and peaceful uses of nuclear energy. "Nuclear" disarmament had supplanted nonproliferation as first in the list.[49] This was not an accident, and while these were not consensus documents, they reflected a growing tendency to accord nuclear disarmament a higher priority than nonproliferation.

This evolution in the characterization of the NPT was clearly influenced by nonaligned countries, which viewed the Bush Administration's emphasis on nonproliferation as excessive and coming at the expense of nuclear disarmament. This perception was reinforced by its reversal of many of the Clinton Administration's Article VI policies. Nonaligned statements in 2007 and 2008 stressed that the lack of balance in implementation of the three pillars "threatens to unravel the NPT regime."[50] The Bush Administration did not em-

---

ence of the Parties to the Treaty on the Non-Proliferation of Nuclear Weapons, May 10, 2004, NPT/CONF.2005/PC.III/WP.27, p. 1, available from *www.un.org/ga/search/view_doc.asp?symbol=NPT/CONF.2005/PC.III/WP.27*.

49. Chairman's working paper, Preparatory Committee for the 2010 Review Conference of the Parties to the Treaty on the Non-Proliferation of Nuclear Weapons, May 11, 2007, NPT/CONF.2010/PC.I/WP.78, p. 1, available from *https://documents-dds-ny.un.org/doc/UNDOC/GEN/N07/353/04/PDF/N0735304.pdf?OpenElement*; and Chairman's Working Paper, Preparatory Committee for the 2010 Review Conference of the Parties to the Treaty on the Non-Proliferation of Nuclear Weapons, May 9, 2008, NPT/CONF.2010/PC.II/WP.43, p. 1. available from *https://documents-dds-ny.un.org/doc/UNDOC/GEN/N07/353/04/PDF/N0735304.pdf?OpenElement*.

50. Statement by Cuba on behalf of the Group of Non-Aligned States Parties to the Treaty of the Nonproliferation of Nuclear Weapons, General Debate of the First Session of the Preparatory Committee for 2010 Review Conference of the

brace pillars as its official mantra, but did not object either—while making clear that nonproliferation was the pillar deserving priority attention.[51]

The Obama Administration's first mention of the NPT as having three pillars was on May 5, 2009—one month after the Prague speech—in then-Assistant Secretary of State Rose Gottemoeller's opening U.S. statement to the third session of the Preparatory Committee for the 2010 Conference.[52] In outlining U.S. views on the three pillars, Gottemoeller led with disarmament, as did the President in Prague when describing the NPT's basic bargain. In documents and speeches over the next year, the administration fleshed out its view on the mutually reinforcing nature of the pillars.

In a March 5, 2010, statement on the 40th anniversary of the NPT's entry into force President Obama said: "Each of these three pillars—disarmament, nonproliferation and peaceful uses—are [sic]

States Parties to the Treaty on the Non-Proliferation of Nuclear Weapons, Vienna, April 30, 2007, p. 2, available from *www.reachingcriticalwill.org/images/documents/Disarmament-fora/npt/prepcom07/statements/30aprilNAM.pdf*; and Statement by Indonesia on behalf of the Group of Non-Aligned States Parties to the Treaty of the Nonproliferation of Nuclear Weapons, General Debate of the Second Session of the Preparatory Committee for 2010 Review Conference of the States Parties to the Treaty on the Non-Proliferation of Nuclear Weapons, Geneva, April 28, 2008, p. 2, available from *www.reachingcriticalwill.org/images/documents/Disarmament-fora/npt/prepcom08/statements/NAMApril28.pdf*.

51. John Wolf, Remarks to Second Session of the Preparatory Committee for the 2005 Review Conference of the Parties to the Treaty on the Non-Proliferation of Nuclear Weapons, Geneva, Switzerland, April 28, 2003, p. 6, available from *www.reachingcriticalwill.org/images/documents/Disarmament-fora/npt/prepcom03/2003statements/28April_US.pdf*; and Christopher A. Ford, "The 2010 NPT Review Cycle So Far: A View from the United States of America," Wilton Park, UK, December 20, 2007, available from *https://2001-2009.state.gov/t/isn/rls/rm/98382.htm*.

52. Rose Gottemoeller, "Opening Statement at the Third Session of the Preparatory Committee for the 2010 Nuclear Non-Proliferation Treaty Review Conference," UN Headquarters, New York, May 5, 2009, available from *https://2009-2017-usun.state.gov/remarks/4358*.

central to the vision that I outlined in Prague of stopping the spread of nuclear weapons and seeking a world without them."[53]

As the Conference approached, Under Secretary of State Ellen Tauscher underlined the need to strengthen the NPT's three pillars and said "this review conference is different from past conferences." She went on to highlight Secretary of State Hillary Clinton's decision to lead the U.S. delegation, in contrast to the previous administration's delegating that task in 2005 to an Assistant Secretary of State. Tauscher pointed to North Korean and Iranian actions as evidence "that the nuclear nonproliferation regime is under great stress and is fraying at the seams."[54]

Secretary Clinton delivered the major U.S. address to the Conference on May 3, 2010. She said the stakes were high and that we could be confronted with a new wave of proliferation unless all parties fulfilled their responsibilities under the NPT. To prevent that possibility "we must commit ourselves to strengthening the three pillars of the nonproliferation regime."[55]

A review of statements by U.S. officials through 2015 reveals no change in this approach. At the UN General Assembly on October 7, 2014, Gottemoeller, now the Under Secretary of State for Arms Control and International Security, stated that "it is important to focus on all three pillars of the NPT" and that "NPT pillars are mutu-

---

53. "Statement by President Barack Obama on the 40th Anniversary of the Nuclear Nonproliferation Treaty," March 5, 2010, available from *https://geneva.usmission.gov/2010/03/08/obamanuclear-nonproliferation-treaty/*.

54. Ellen Tauscher, "Previewing the NPT Review Conference," Remarks at the Center for American Progress, Washington, DC, April 29, 2010, available from *www.state.gov/t/us/141029.htm*; and Ellen Tauscher, Susan Rice, and Susan Burk, "Briefing on the Nonproliferation Treaty Review Conference" Press Conference, Washington, DC, April 30, 2010, available from *https://2009-2017.state.gov/t/us/141029.htm*.

55. Hillary Clinton, "Remarks at the Review Conference of the Nuclear Nonproliferation Treaty," United Nations, New York, May 3, 2010, available from *https://www.state.gov/secretary/travel/rm/2010/05/141424.htm#*.

ally reinforcing and implementation of each is a shared responsibility." At the NPT Review Conference on April 27, 2015, Secretary of State Kerry stated that each pillar "is an essential ingredient to the full embodiment of the NPT. The NPT cannot stand unless all three of those pillars are study enough to support it."[56] At a conference in Japan on November 1, 2015, Deputy Assistant Secretary of State Anita Friedt paid tribute to the "grand bargain" of the NPT (nuclear disarmament, nonproliferation, peaceful uses) and said "All state parties to the NPT value the treaty's three mutually reinforcing pillars that remain strong even in the face of periodic setbacks."[57]

Why the U.S. Embrace of the Three Pillars Approach?

The first order question is the easiest to answer: Does the three pillars strategy suggest a different legal interpretation of the NPT's main obligations? In short, the answer is no. While the President elevated the peaceful uses and nuclear disarmament components of the treaty, and even placed nuclear disarmament as first among the three, the actual language used to characterize the pillars is consistent with the treaty's undertakings.[58] Moreover, other NPT statements by the Administration during and since the 2010 Conference continue to use a long-standing consensus formulation that creates no ambiguity about nonproliferation as the treaty's central purpose

---

56. Remarks by Secretary of State John Kerry at the 2015 Nuclear Nonproliferation Treaty Review Conference, April 27, 2015, available from *http://www.cfr.org/nonproliferation-arms-control-and-disarmament/remarks-secretary-kerry-2015-nuclear-nonproliferation-treaty-review-conference/p36497*.

57. Remarks by Anita Friedt, Principal Deputy Assistant Secretary of State, Bureau of Arms Control, Verification and Compliance before the 61st Pugwash Conference on Science and World Affairs, Nagasaki, Japan, November 1, 2015, available from *https://2009-2017.state.gov/t/avc/rls/2015/249252.htm*.

58. "Remarks by President Barack Obama," Prague; and "Statement by President Barack Obama on the 40th Anniversary of the Nuclear Nonproliferation Treaty."

and that it underpins the other pillars. There are several variations, but the substance of these statements cites the NPT's role in preventing the spread of nuclear weapons and/or as the cornerstone of the nonproliferation regime; while noting that the treaty also provides an essential foundation for progress/negotiations on disarmament and for cooperation in the peaceful uses of nuclear energy.[59]

The most plausible explanation for the three pillars strategy is that the administration wished to convey *equal political* weight to all three components of the treaty. The United States saw the NPT as at serious risk due to proliferation concerns and to perceptions that the nuclear powers were not serious about Article VI. This situation had been exacerbated by the previous administration's tendency to focus far more on the treaty's nonproliferation mission than on disarmament. Embracing the NPT as three reinforcing pillars and highlighting the reenergized U.S. commitment to nuclear disarmament was seen as a credible way to invigorate NPT implementation and restore U.S. leadership.

This approach also reflects a belief that an unequivocal U.S. commitment to all aspects of the treaty, including Articles IV and VI, is essential to the long run success of the NPT's nonproliferation mis-

---

59. Examples include "United States information pertaining to the Treaty on the Non-Proliferation of Nuclear Weapons," submitted to the 2010 Review Conference of the Parties to the Treaty on the Non-Proliferation of Nuclear Weapons, NPT/CONF.2010.45, May 7, 2010, p. 1, available from *www.un.org/ga/search/view_doc.asp?symbol=NPT/CONF.2010/45*; Rose Gottemoeller, "Third Meeting of the Preparatory Committee to the 2015 Nuclear Non-Proliferation Treaty Review Conference, New York, April 29, 2014, p. 1, available from *http://iipdigital.usembassy.gov/st/english/texttrans/2014/04/20140429298471.html#axzz4XYKTJZrX*. P5 statements also routinely include this characterization: See for example "Joint Statement on First P-5 Follow-Up Meeting to the NPT Review Conference," July 1, 2011, available from *https://geneva.usmission.gov/2011/07/05/p-5-npt-review-conference/*; and "P-5 Statement to the 2012 NPT Preparatory Committee," May 3, 2012, p.1, available from *www.un.org/ga/search/view_doc.asp?symbol=NPT/CONF.2015/PC.I/12* and "Joint Statement by P-5 at the 2015 NPT Review Conference," April 30, 2015, p.1,available from *http://www.un.org/en/conf/npt/2015/statements/pdf/P5_en.pdf*.

sion. This belief is met with skepticism by some,[60] but others view credible implementation of Article VI as fundamental to sustaining the NPT's value as a barrier to proliferation.[61] The reality, of course, is never quite as simple as frequently asserted.[62] It is important not to overstate the linkage; certainly the violations uncovered over the past 25 years in Iraq, North Korea, Iran, Libya, and Syria cannot be traced to compliance or non-compliance with Article VI. Moreover, the reluctance of some non-nuclear states to hedge on their support for nonproliferation initiatives in recent years cannot be sourced solely to Article VI. Still, the undertakings in Articles IV and VI are legally-binding and manifest the political consensus that led to the NPT in the 1960s. To downplay their importance or appear condescending to non-nuclear states' criticism of their implementation risks weakening support for the treaty as a whole.

A push on Article VI was a natural fit for the Obama Administration. The President believes in the benefits of observing and enforcing international norms, including the NPT, and in the importance of the United States meeting its obligations pursuant to such norms.

---

60. Christopher A. Ford offers a thorough critique, "Nuclear Disarmament, Nonproliferation, and the 'Credibility Thesis,'" September 2009, Briefing Paper, Hudson Institute, pp. 8-14, available from *02e18f7.netsolhost.com/New_Paradigms_Forum/Documents_and_Links_files/Ford%20Hudson%20Paper%20on%20Credibility.pdf*.

61. Scott D. Sagan and Jane Vaynman: "Conclusion: Lessons Learned from the 2010 Nuclear Posture Review," *Nonproliferation Review*, Vol. 18, No. 1, March 2011, p. 238; Lewis A. Dunn, "The NPT: Assessing the Past, Building the Future," Nonproliferation Review, Vol. 16, No. 2, July 2009, pp. 163-164; and Lewis A. Dunn, ed., "Foreign Perceptions of U.S. Nuclear Policy and Posture: Insights, Issues and Implications," report prepared by Science Applications International Corporation for the Advanced Systems and Concept Office, Defense Threat Reduction Agency, December 12, 2006.

62. Henry Sokolski, *Underestimated: Our Not So Peaceful Nuclear Future,* Carlisle, PA: Strategic Studies Institute, 2015, pp. 7-46, available from *www.npolicy.org/thebook.php?bid=34*. Sokolski presents differing views on the relationship between nuclear weapons reductions and nonproliferation and finds them all wanting to various degrees.

The President had campaigned in favor of a more aggressive U.S. role in reducing reliance on nuclear weapons. By the time of the review conference, these policies had been put into place and the New START Treaty had been signed. The previous administration's opposition to treaty-based nuclear arms control was seen as having created a trust deficit for the United States, which the new administration was determined to reverse.

A strong emphasis on the peaceful use "pillar" also made sense. Iran was going to make Article IV a major issue in any event, claiming that its "rights" to a peaceful nuclear program were being illegally curtailed. The administration was in a good position to counter these claims by citing Article IV's affirmation that peaceful uses must conform with the NPT's nonproliferation undertakings and by reinforcing the President's Prague policy that those who break the rules should face consequences. Moreover, regardless of Iranian protests, the United States has a long and credible history of effective implementation of Article IV, which the Administration planned to buttress at the 2010 Conference by a "peaceful uses" initiative targeted at developing countries.

In sum: The administration viewed the NPT at a crossroads and circumstances suggested that a heightened U.S. emphasis on Articles IV and VI at the 2010 Conference would restore a sense of stability to its implementation. The result was an embrace of the basic bargain concept of the NPT and its characterization as comprising three pillars that are mutually reinforcing and interrelated. This nomenclature had been used in meetings of the Preparatory Committee for the Conference that were held before the Obama Administration came into office. It probably seemed unexceptional and compatible with a desire to restore U.S. bona fides in regard to Article VI. While this strategy represents a substantial increase in U.S. political support for equal treatment among nonproliferation, disarmament, and peaceful uses, it did not alter the treaty's basic undertakings or represent any change in the object and purpose of the NPT. One can question the strategy, but the substance of the administration's position provides no persuasive evidence of a change in treaty interpretation.

The Risks

The three pillars strategy, along with the prominence given to the Article VI pillar, strays from the traditional U.S. portrayal of the NPT. It has downsides, and the following discussion attempts to illuminate them. The risks may not become serious or unmanageable, and would not likely be manifest except over an extended period. But they are not trivial, and future U.S. NPT diplomacy must prevent these risks from growing and undermining the treaty.

The bulk of NPT non-nuclear states, represented by nonaligned countries, have long pushed for more focus on Article VI. From 2003-2008 this nonaligned critique was compounded by a perception that the Bush Administration was overemphasizing nonproliferation while dialing back on Article VI. Fair or not, this view led some nonaligned countries to suggest that support for stronger nonproliferation measures should be withheld in the absence of progress on Article VI. Another point frequently heard is that work on the pillars should be balanced and/or advanced together. While it sounds benign, acceptance of this linkage gives NPT non-nuclear states a "pass" on bolstering nonproliferation if they are dissatisfied with Article VI implementation. A similar example lies in the tendency of some states to have downplayed Iran's violations while trumpeting allegations of nuclear weapon state failure to comply with Article VI.

While the reasons for these nonaligned positions are more complex than simple dissatisfaction with nuclear weapon state performance under the NPT, it is not in the interests of the treaty for this situation to persist. Such attitudes can slowly erode the barriers to the NPT's nonproliferation mission and should be strongly countered. Yet, by accepting the three pillars of the NPT as politically equivalent and mutual reinforcing, the Administration appears to have legitimized these linkages. There is also the risk that if the promised Article VI progress does not occur (it takes at least two to conduct good faith negotiations), some NPT non-nuclear states will feel even more justified in their lukewarm support for nonproliferation.

Finally, even assuming progress resumes on Article VI, it is inevitable that the nuclear disarmament process will face difficulties and occasionally stall as nuclear stockpiles go lower and the other nuclear powers join the negotiations. One need only consider the task of preserving security at ever lower levels of nuclear weapons among disparate states amid an unpredictable security environment to realize the enormity of the challenge. During this period, the NPT will need to hold the line against further proliferation. That could be more difficult if treaty parties have come to view nonproliferation and nuclear disarmament as inextricably linked. In general, legitimizing the NPT as a bargain between nuclear weapon states "moving toward nuclear disarmament" and non-nuclear states "forsaking nuclear weapons" invites a quid pro quo mentality toward advancing these goals—when in reality each is worthy of pursuit independently.

Theoretically, the same negative dynamic might operate for Article IV, i.e. granting equivalency to the peaceful use pillar could encourage more conditional linkages to nonproliferation by NPT parties who feel their rights to certain peaceful nuclear facilities are being curtailed. This dynamic already exists, but it does not carry the same weight as Article VI. The nonaligned Article IV critique over the "right to access" applies primarily to enrichment and reprocessing technology, which for virtually all NPT non-nuclear parties is not an economically viable fuel cycle option. Thus, their interests are not directly affected.

Many non-nuclear states' complaints about Article IV are perfunctory and would not likely become more substantive even if peaceful uses acquired a status under the NPT equivalent to nonproliferation. (In contrast, most of these states were persuaded long ago that nuclear disarmament occupies a level of importance equal to that of nonproliferation.) It also seems unlikely that "pillar equivalence" in this area would have any impact on weakening supplier policies; restraint on certain nuclear technologies and ever stricter export controls have a long pedigree. Moreover, several nonaligned states are now associated with such controls by having joined the suppliers group. In general, elevating peaceful use to the status of nonproliferation under the NPT would appear to have a far lower risk to the NPT's effectiveness than doing the same for nuclear disarmament.

A similar, but more subtle, risk is that characterizing the NPT as three inter-related and mutually reinforcing pillars conveys a false sense of equivalency in terms of meeting the NPT's object and purpose. Taken to an extreme, along with the image of three pillars holding up the NPT, it implies that a failure or weakening in any one of the pillars would have a direct and equal effect on the stability of the treaty. Thinking of the NPT in this way misreads what the Treaty was designed to accomplish and promises more than it can deliver.

It is axiomatic that the failure to prevent even one NPT non-nuclear weapon state from acquiring nuclear weapons would further complicate nuclear disarmament. Indeed, the NPT was regularly identified during the negotiations *as a step toward nuclear disarmament*—an admission that nonproliferation is necessary to achieve that goal. Moreover, while the NPT is not necessary to the pursuit of nuclear energy for peaceful purposes, the treaty has fostered a higher degree of confidence in peaceful uses than might have otherwise have been the case; and Article IV has provided useful leverage for developing countries to gain access to such benefits. The parties themselves have regularly described the NPT as the foundation for pursuit of nuclear disarmament and of peaceful uses. In other words, its widespread success in codifying the norm of nonproliferation has helped to preserve the 70-year goal of eliminating nuclear weapons and the promise of peaceful uses of nuclear energy for all nations. But to suggest a mutual degree of interdependence in the other direction is misleading and continued emphasis on it would weaken the treaty in the long run.

Recall that the NPT nonproliferation mission continued to grow in popularity during the worst times of the Cold War when little progress was recorded on the "good faith" negotiations called for in Article VI and deployed strategic weapons increased to 10,000 on each side. Had the NPT been evaluated in 1990 on the basis of "three mutually-reinforcing pillars," it would have been judged a failure given the meager record of Article VI implementation. Conversely, the substantial decrease in nuclear weapons stockpiles over the past 25 years has been met with a growth in nuclear proliferation and violations of the NPT's nonproliferation undertakings. This trend is the opposite of what one would expect if progress in Article VI implementation enabled nonproliferation. Similarly, to suggest that peaceful nuclear use under Article IV reinforces Articles I, II, and III is unpersuasive since it is only through conformance with these nonproliferation undertakings that a state's nuclear program can earn its status as peaceful in the first place.

Another factor relevant to this discussion of equivalency among the so-called pillars is the actual language of the treaty. How explicit are the undertakings and can they be measured? There are stark differences: The nonproliferation undertakings establish a prohibition on the acquisition and transfer of nuclear weapons and require non-nuclear powers to accept verification by the IAEA. In contrast, Article IV requires parties "to facilitate...the fullest possible exchange" of peaceful nuclear commodities and Article VI requires parties to "negotiate in good faith ...on effective measures" related to disarmament. The nonproliferation obligations are far more explicit. Of course, that was not an accident; the treaty was created to prevent the spread of nuclear weapons to additional countries. Nonperformance by NPT parties of Articles II and III can be measured, if not early in the process of acquiring nuclear weapons then more likely once the prohibited activities involve the use of nuclear material and certainly once a nuclear device is detonated. However, how is compliance with Articles IV and VI measured? Persistent charges of violations of these Articles suggest that some NPT parties have set their own standards; but the treaty offers no guidance. That too is not an accident, as the drafters deliberately formulated the Article IV and VI undertakings to encompass general obligations rather than specific measures. Positing equal value of the three pillars would make sense only if the undertakings were equally explicit, making compliance with each equally measurable and enforceable. That is not the case.

Certainly, a good record of fulfilling the general obligations of Articles IV and VI maximizes the perceived benefits for non-nuclear states and helps to sustain the NPT and its central purpose of non-proliferation. In that regard, the NPT is no different than any other multilateral treaty. Its core obligations are supplemented with ancillary commitments that address the diverse interests of nations whose support is being sought. That doesn't translate into imputing equal importance to each provision in achieving the treaty's purpose.

There is also the danger that viewing the three pillars as of equal political importance could encourage an interpretation that accords them equal legal importance. Continued emphasis on the presumed equality of the pillars could over time lead some NPT parties toward the logic of suspending operation of the treaty because of a claimed material breach of Article VI. A recent revisionist interpretation of the NPT has raised just such a possibility.[63] Admittedly, this is a

---

63. Daniel H. Joyner, *Interpreting the Nuclear Non-Proliferation Treaty*, Oxford: Oxford University Press, 2011; Critics of this approach include Norman A. Wulf, "Misinterpreting the NPT," [Review of Interpreting the Nuclear Non-Proliferation Treaty by Daniel Joyner] *Arms Control Today*, September 2011, available from *legacy.armscontrol.org/2011_09/Misinterpreting_the_NPT*; and Christopher A. Ford, "Misinterpreting the NPT," remarks delivered at a discussion of Dr. Joyner's book organized by the Carnegie Endowment for International Peace, September 30, 2011, available from *www.newparadigmsforum.com/NPFtestsite/?p=1100*.

worst case scenario and current implementation practices of the parties do not suggest that non-nuclear parties are likely to take this step.

In addition to the long-term risks of this strategy, its use has placed the United States at a tactical disadvantage as the obstacles mount to the Prague Article VI agenda. This situation invites the kind of dithering by some non-nuclear parties described above on the NPT's nonproliferation obligations. Leading with Article VI and the three pillars strategy may have yielded short-term benefits in 2010, but continuing to leverage it to advance the NPT offers little promise.

The principle argument against the three pillars strategy is that the NPT was conceived with one pillar, as the title of the treaty makes clear, and can still deliver on only one pillar—nonproliferation. At its core, the benefits of the NPT do not stem from any grand bargain or quid pro quo between the nuclear and non-nuclear powers on nonproliferation, nuclear disarmament, and peaceful uses. Yes, there would be no NPT without Articles IV and VI, and these undertakings must be taken seriously and pursued with diligence, transparency, and due regard for the views of other NPT parties. But it is misleading to assert or imply a relationship of strict interdependence between implementation of Articles IV and/or VI on the one hand and Articles I, II and III on the other. The structure and terms of the NPT created a nonproliferation security pact among its members; the inclusion of Articles IV and VI was never intended to create an obligation equivalent in effect to nonproliferation nor do the terms of these Articles do so.

In its zeal to invigorate the NPT and U.S. leadership, the Obama Administration "double-downed" on nuclear disarmament and the three pillars approach appeared to accept nonproliferation and nuclear disarmament as equal goals under the NPT. Words matter and a continuation of that approach could further exacerbate the tensions inherent in the NPT, which have grown substantially in recent years. Indeed, the lack of progress on the 2009 Prague nuclear agenda almost certainly contributed to the disappointing results of the 2015 NPT Review Conference. By 2015, U.S. leverage due to Article VI had substantially dissipated and the United States was in a weaker position to work out acceptable compromises on unrelated topics that might have led to adoption of a Final Document. On the other hand, the Obama Administration would almost certainly claim that its "rebalancing" act on the NPT helped to preserve and strengthen the coalition pressuring Iran which led to a diplomatic solution shortly after the Review Conference. One could argue both of these points and causal links are virtually impossible to prove. However, the author is not swayed from the general conclusion that in the long run it is not healthy for the United States to pursue a diplomatic strategy that detracts from the NPT's central mission of nuclear nonproliferation.

Strategy Going Forward

The NPT has always drawn bipartisan support, and it is expected that the next Administration—Democrat or Republican—would be prepared to continue efforts to ensure its effectiveness. Implementation of the Iran deal should ensure that the Treaty's nonproliferation mission remains front and center and offers opportunities to strengthen it. But emphatically what should not happen is an effort to renegotiate the 2015 Iran deal or to block its implementation, as some U.S. political leaders have urged. Doing so, even despite good intentions, would undermine U.S. leadership in the nonproliferation field including with the NPT. Reversing widely supported international commitments to which previous Administrations have agreed, such as the nuclear deal, would harm U.S. standing in the world. The ball is in Iran's court to comply scrupulously with the settlement. At this point the only sensible approach would be to closely monitor Iran's compliance and be ready to pounce in the event of a serious violation.

In regard to general NPT diplomacy the United States will not be able to avoid the three pillars terminology. However, it is time to dial back the excessive U.S. focus within the NPT on nuclear disarmament. The notion of the pillars as mutually reinforcing and interrelated should be quietly retired from the U.S. position. Equal attention to implementation of each pillar should be acceptable, but efforts to link progress in one to progress in another should be strongly resisted. It should be enough to stress that the continued value of the NPT depends on respect for and compliance with all three pillars. NPT parties should stop arguing with each other over the relative importance of nonproliferation and nuclear disarmament, and energize efforts to pursue both goals.

To NPT parties that find fault with Article VI implementation by the United States and argue that nuclear disarmament should take priority over nonproliferation, it is important to point out that the increased nuclear weapons risk over the past 25 years is due to proliferation—not from actions by the United States. Major NPT violations were uncovered in North Korea, Iraq, Iran, Libya and Syria —infractions that led to increased tensions and insecurities. The NPT's nonproliferation undertakings were instrumental in mounting challenges to these violations. Strengthening IAEA safeguards, seeking global acceptance of the Additional Protocol, and ensuring consequences for violators are all worthy of support in advancing this mission. Without efforts to hold the line against NPT violators, the vision of a nuclear weapons free world would become ever more distant. Just because the NPT is nearly universal and most parties honor their obligations is no reason for complacency.

While reinforcing the NPT's nonproliferation purpose, the United States should also ensure that it maintains a credible record of meeting U.S. obligations under Articles IV and VI. Along these lines, there seems little cost in retaining the U.S. commitment to seek the peace and security of a world without nuclear weapons, while making clear that the United States will retain nuclear weapons as long as others have them. If presidents as disparate as Ronald Reagan and Barack Obama can make such a pledge, it should not be difficult for the president who takes office in January 2017 to do the same.

Despite obvious obstacles to progress on nuclear disarmament, the United States should recognize that examining ways to achieve substantial further reductions of nuclear weapons and to do so in a verifiable manner are now in the mainstream. A rejection of global nuclear disarmament as a desirable goal or of activities that advance this goal would brand the United States as an outlier and severely erode U.S. influence on nonproliferation. Despite the passage of seven decades and a history of non-use, the impulse to eliminate the world's most devastating weapons remains strong among governments around the world and increasingly by civil society. There is nothing inherently incompatible with policies that sustain a credible focus on the goals of Article VI while pursuing a posture that recognizes the essential albeit declining role of nuclear weapons in U.S. national security.

The NPT has proved its worth in confronting nuclear proliferation. While established during the Cold War and approaching its 50th anniversary, the Treaty has demonstrated its continued relevance in the highly unpredictable political and security environment of the 21st century. U.S. leadership made the NPT possible and has ensured its continued vitality over the years. The next Administration is well positioned to continue this record and to do so in an environment that is conducive to a balanced approach to U.S. NPT diplomacy and a well-timed focus on nonproliferation.

CHAPTER 3

Safeguards and the NPT: Where Our Current Problems Began

Leonard Weiss

Introduction

The advent of the nuclear age spawned the creation of an international system intended to spread the benefits of nuclear technology while preventing the spread of the bomb. The system includes organizations and rules that have evolved in an attempt to realize this difficult dual mission, but it has failed to prevent a growing number of nations from either producing nuclear weapons or putting themselves in a position to produce such weapons quickly. The nature of current commercial nuclear technology makes it feasible for a country to use a civilian nuclear power program to shield a weapons development program. This is not a surprise. The security and proliferation problems visible today in the nuclear area were foreseen by some scientists, but the extravagant visions of nuclear nirvana blinded the champions of nuclear energy to the fact that spreading the technology would allow more and more countries to increase their nuclear "latency" (their closeness to nuclear weapons capability). It is instructive to examine some early milestones in the mitigation of nuclear proliferation risks that unintentionally led to current proliferation problems and still have the potential to promote further increases in nuclear "latency."

Atoms for Peace and the Cold War

The first successful controlled nuclear chain reaction, on December 2, 1942, not only brought the prospect of nuclear weapons closer to reality, but also the prospect of controlled nuclear energy for civilian purposes. The latter idea was set aside while World War II raged onward, but once the bombs were dropped on Hiroshima and Nagasaki and the war ended, the question of the future of nuclear energy, both military and civilian, was engaged. The Cold War began even before British Prime Minister Winston Churchill's famous "Iron Curtain" speech[1] at Fulton, Missouri in 1946, and the nuclear issue was as central to it as the question of post-war Soviet domination of Eastern Europe. Indeed, the issues were linked in the eyes of some policymakers, but there was hope that a post-war U.S.-Soviet confrontation could be avoided, with the internationalization of nuclear energy as the linchpin allowing the continuation of the wartime alliance between the two nations. Alas, it was not to be. Between Soviet Premier Joseph Stalin's distrust of the West—based in part on the history of the failed Western expeditionary forces sent to crush the Russian Revolution—and the abject fear of Communist ideology by Western reactionaries like U.S. Secretary of State James Byrnes, who saw the bomb as a potential instrument of coercion against the Soviets, there was little likelihood that the bomb would be anything other than an area of intense competition once the Soviets inevitably began constructing their own nuclear arsenal.

The group of scientists who designed and developed the bomb included many who saw nuclear energy as a potential boon to mankind that would eventually overshadow its provenance as a weapon of mass destruction and many who thought that international cooperation on development could make that happen. But the reality of a U.S. monopoly on the weapon, short-lived though it was destined to be, was too enticing to the United States and other Western pow-

---

1. Winston Churchill, "Sinews of Peace," Speech at Westminster College, Fulton, Missouri, March 5, 1946, available from *www.historyplace.com/speeches/ironcurtain.htm*.

ers and made the prospect of a U.S.-Soviet entente extremely difficult if not impossible. Nonetheless, even as the Soviets worked feverishly to build their first nuclear device, the United States was giving serious thought to the question of how international development of nuclear energy for peaceful purposes could be carried out. The question was given to a committee set up by then-Assistant Secretary of State Dean Acheson and chaired by David Lilienthal, the head of the Tennessee Valley Authority (TVA). Lilienthal appointed an advisory panel containing Robert Oppenheimer to do the analysis. The committee's report,[2] issued on March 28, 1946, was sobering. In essence, the committee said that the only way of developing nuclear energy for peaceful purposes without spreading bomb technology was to prohibit all national development and provide international ownership of and responsibility for all nuclear materials and advanced technology.

The Acheson-Lilienthal report did not sit well with the guardians of free enterprise in the West or with the Soviets, who wanted the bomb, so the notion of international ownership and control was scrapped and national nuclear rivalries began their rise. The U.S. nuclear monopoly was institutionalized by Congress via the enactment of the Atomic Energy Act of 1946, which formally declared U.S. government ownership of all nuclear materials under its control and placed research, development, and demonstration of nuclear energy in the hands of a new agency, the Atomic Energy Commission (AEC)—a five member board appointed by the President with the advice and consent of the Senate. A Senate-House committee, the Joint Committee on Atomic Energy (JCAE), was established to provide legislative and investigatory oversight over the nuclear enterprise, and strict rules were established limiting the sharing of nuclear information and technology, with severe penalties for violations.

---

2. *A Report on the International Control of Atomic Energy* (The Acheson-Lilienthal Report), report prepared for the Secretary of State's Committee on Atomic Energy, Washington, DC: Government Printing Office, 1946, available from *www.learnworld.com/ZNW/LWText.Acheson-Lilienthal.html*.

But U.S. secrecy did not prevent the Soviets from exploding a fission device of their own in 1949 and a thermonuclear device in 1953. When Dwight Eisenhower became President in January 1953 he was presented with a stark report about growing Soviet nuclear and conventional capabilities and a proposal to inform the U.S. public of the dangers of nuclear war.[3] The report reflected the fear that without some restraint in the arms race the Union of Soviet Socialist Republics (USSR) might achieve a nuclear force level (believed by the authors to be about 600 fission weapons with a 40 kiloton yield) that, if dropped on the United States, could be sufficient to destroy the U.S. economy beyond recovery in a few years—a "knockout blow" in the parlance of the report.[4] Even if the U.S. had more weapons, the report argued, the Soviets might be tempted to attack if they believed a knockout blow was feasible. The panel's advice to Eisenhower was to inform the American people of the growing danger in order to obtain support for nuclear arms control, i.e., support for reducing and keeping the stockpile of weapons on both sides below the point where a knockout blow would be seen as not possible. This created a policy dilemma for Eisenhower, who intended to carry out a new program making nuclear weapons a central part of a greatly expanded military buildup. At the same time, he recognized the difficulty of negotiating an arms control agreement with the Soviets, which would require an intrusive inspection system that the national security apparatus of the USSR was predisposed to oppose.

President Eisenhower finessed this problem by seizing on the idea of the superpowers reducing, if not their weapons, at least their stockpile of fissile materials by contributing a certain amount for peaceful purposes. His hope was that the Soviets might find it difficult to match a U.S. contribution, in which case either the

---

3. Leonard Weiss, "Atoms for Peace," *Bulletin of the Atomic Scientists*, Vol. 59, No. 6, November 2003, pp. 33-44.

4. Henry Sokolski, *Best of Intentions: America's Campaign Against Nuclear Weapons Proliferation*, Westport, CT: Praeger, 2001, p. 26.

number of future USSR weapons would be reduced or the United States would gain a propaganda victory. This idea initially generated skepticism from the defense establishment, which saw it as possibly compromising the "new look" defense posture designed to increase U.S. reliance on nuclear weapons, but some advisors, including AEC Chairman Lewis Strauss, became supporters of the idea. Strauss foresaw the possibility of growing fissile material contributions for peaceful purposes becoming so large over time that a military fissile material production cutoff agreement might be possible.[5] While protection of contributed material from theft was discussed, the danger of horizontal proliferation as a security threat once the contributed materials were distributed was given short shrift. This occurred partly because of the quaint notion extant that a country having only a small stockpile of weapons was not a security threat[6] and mostly because of the apparent strategic and propaganda advantage the proposal provided the Americans over the Soviets. Accordingly, in December 1953, President Eisenhower went before the United Nations (UN) and laid out an attractive vision of peaceful nuclear development in which all nations could participate via a nuclear fuel bank into which the United States and the USSR would contribute equal amounts. The bank was to be administered by an international atomic energy agency, but the question of how the bank would operate to ensure peaceful use of the material was deferred. The Soviets were not blind to the proliferation consequences of Eisenhower's proposal. After President Eisenhower's speech, Soviet Foreign Minister Vyacheslov Molotov reportedly told U.S. Secretary of State John Foster Dulles that the proposal would spread bomb materials worldwide.[7]

Meanwhile, President Eisenhower's vision triggered a demand by

---

5. Ibid, p. 29.

6. Ibid, p. 32.

7. Paul Leventhal, "Fixing Ike's Flawed Nuclear Plan," *Nuclear Control Institute*, December 8, 2003, available from www.nci.org/06nci/10/Fixing%20Ike%27s%20Flawed%20Nuclear%20Plan.htm.

Congress for an expanded program of nuclear development that industry contractors had been seeking. This required new legislation, and the Atomic Energy Act of 1954 (AEA) was enacted, which gave industry the ability to own nuclear materials and develop nuclear technology. Under the AEA, the United States began what was called an "Atoms for Peace" program involving the building of research reactors, fueled with highly enriched uranium (HEU), for export to many countries in the hope that this would lead to larger contracts for American companies as nuclear technology became commercialized.

Recognizing and Ignoring Safeguards Issues

The loosening of government controls was accelerated when the United States proposed an international scientific conference that took place in Geneva in 1955, which was designed to push nuclear development worldwide, including the idea of a uranium bank.[8] The Soviets were not expected to participate in the conference. As indicated earlier, they had criticized the bank proposal—along with the notion of spreading nuclear technology for peaceful purposes—on the grounds that it could make the acquisition of nuclear weapons materials available to more countries. Americans were not entirely blind to this concern and required that spent fuel from research reactors provided by the United States be returned to the United States for reprocessing. But that was as far as any thoughts about safeguards went. However, when the Soviets, in a surprise move, agreed to participate in the conference and the development of the fuel bank, the American delegation—whose scientific head was I.I. Rabi—scrambled to come up with a proposal, for the Soviets to consider, on safeguarding the nuclear materials to be distributed by the proposed International Atomic Energy Agency (IAEA). At that point the IAEA did not have a recognized charter, which had to

---

8. Richard Hewlett and Jack Holl, *Atoms for Peace and War, 1953-1961*, Berkeley: University of California Press, 1989, available from *energy.gov/sites/prod/files/2013/08/f2/HewlettandHollAtomsforPeaceandWarComplete.pdf*.

be negotiated and was not formally adopted by the UN until 1957.

Based on a late night discussion within the American delegation in their Geneva hotel, Rabi presented an ad hoc proposal to the Soviet delegation to safeguard weapons materials from theft or diversion by tagging the material with a highly radioactive element, uranium-232 (U-232), so that the path of the material could be traced from fabrication to reprocessing. The head of the Soviet delegation, Dmitri Skobeltsyn, along with his colleagues, studied the tagging proposal and then challenged it by pointing out that the decay chain of U-232 had a "dead period" during which it produced the daughter element thorium-228. The latter element was not a hard gamma emitter and thus presented no serious safety problem if chemically separated out at that point in the decay chain. It was only when the chain reached radium-224 that a sufficiently energetic gamma emission was produced that would provide protection against diversion of the material.[9] The Soviets concluded that "spiking" was not an effective safeguards measure, and the Americans realized they had to go back to the drawing board.

Gerard Smith, the diplomatic head of the American delegation, was so chagrined by the ease with which the U.S. proposal was dismissed that he returned to Washington and raised the question of whether the United States should slow down its plans for nuclear transfers until safeguards were better understood, effective, and deployed. The answer by John A. Hall, the Director of the AEC Division of International Affairs, backed up by AEC Commissioner Willard Libby and Chairman Lewis Strauss, was an emphatic "no."[10] A safeguards system that would be fully effective was deemed too costly and too intrusive to be acceptable to nuclear recipients. The U.S. policy became "sell now, do safeguards later." However, it was agreed that an engineering study of safeguards was needed, so an AEC contract was issued to the Vitro Corporation, which pro-

---

9. Ibid., p. 314.

10. Ibid., p. 317.

duced a report in September 1956 stating that even with a 90% probability of detecting an unauthorized diversion of nuclear materials, one could divert a bomb's worth of plutonium within 5 years without detection. That meant that an effective safeguards system would have to contain political and diplomatic elements, as well as technical ones. A task force created by the AEC to study policy issues regarding reactor development in the United States and abroad produced a report that said Atoms for Peace might be an engine for the proliferation of nuclear weapons to underdeveloped or small countries. Moreover, maximum assurance against diversion would require access to all facilities, areas, and records of the country, as well as intrusive methods of surveillance. But the task force did not consider diversion as the main proliferation concern. The greatest threat, they believed, came from the training of nuclear scientists and engineers in reactor technology and advanced nuclear technology such as reprocessing.[11] The idea that diversion was not a serious national security problem was shared by elements within the Eisenhower administration, including Special Representative for Disarmament Harold Stassen, who stated his belief in 1955 that small nuclear arsenals by non-weapons states were not a national security threat to the United States and could act "as an essential counterpoise to a growing USSR nuclear weapons threat."[12] This was echoed by high level personnel of the Department of Defense who believed proliferation could even work to U.S. advantage if France and Japan became weapons states while the Soviets restrained their unreliable allies in Eastern Europe.[13] Later on, Stas-

---

11. Ibid.

12. Progress Report Prepared by the President's Special Assistant (Stassen), May 26, 1955, *Foreign Relations of the United States*, 1955-1957, Volume XX, Regulation of Armaments; Atomic Energy, Document 33, available from *history.state. gov/historicaldocuments/frus1955-57v20/d33*.

13. See, Shane Maddock, "The Fourth Country Problem: Eisenhower's Nuclear Nonproliferation Policy," *Center for the Study of the Presidency*, 1998, available from *www.thefreelibrary.com/The+Fourth+Country+Problem%3a+Eisenhowe r%27s+Nuclear+Nonproliferation...-a053390297*. Also see, Memorandum for

sen elaborated his view, now totally discredited by historical fact, that the existence of nuclear weapons states "on various sides" was a deterrent to diversion of nuclear materials and the creation of more states with nuclear weapons.[14]

In 1956, negotiations on the IAEA charter began in Vienna, and U.S. diplomatic corps entrusted with the negotiations approached them with a more realistic notion of the serious national security risks posed by nuclear diversion. The head of the U.S. delegation to the negotiations, Ambassador James Wadsworth, recognized the security problems created by Atoms for Peace. In a speech before the Conference on the Statute of the IAEA on October 15, 1956, Wadsworth explained the need for safeguards and the danger of nuclear development without them:

> Small amounts of material used or produced in the course of agency-supported peaceful projects can be adapted for use in weapons of destructive force almost beyond comparison with the most powerful weapons of the pre-atomic era, and more important still is the possibility that the explosion of only one such weapon in a local conflict might be enough to set off a worldwide conflagration.[15]

This notion was underscored during the so-called Yom Kippur War

---

the Secretary of Defense, November 18, 1955, Document 219, NSA, Non-Proliferation; Informal Notes of a Meeting of the National Security Council Planning Board, Washington, DC, December 21, 1955, 10 a.m.-12:30 p.m., *FRUS*, 1955-1957, Vol. 20, pp. 245-250; Memorandum of Conversation among the President's Special Assistant (Stassen) and the Joint Chiefs of Staff, Pentagon, Washington, DC, January 24, 1956, 4 p.m., *FRUS*, 1955-1957, Vol. 20, pp. 276-279.

14. As quoted in Sokolski, p. 32. See also, "Statement by the United States Representative (Stassen) to the Disarmament Subcommittee: Nuclear Weapons and Testing, March 20, 1957," in U.S. Department of State, *Documents on Disarmament 1945-1959*, vol. 2, pp. 765-68.

15. See, U.S. Department of State, *Department of State Bulletin*, Vol. 35, November 19, 1956, p. 816.

between Israel and Egypt in 1973 when Israel, which at one point was losing the war, had ostensibly armed its small nuclear arsenal and would have used the weapons had it faced an existential collapse.[16] U.S. resupply of conventional weapons to Israel helped turn the military situation toward Israel's favor, and the Israelis were persuaded to show some restraint to ensure that the feared intervention on the Egyptian side by the Soviets did not materialize. The full story of these events is still classified, but all sides understood that the use of even one nuclear weapon by the Israelis could have had a catalytic effect in raising the threat of nuclear war between the United States and the USSR.

At the same time Ambassador Wadsworth was persuading the negotiators on the IAEA charter to adopt the requirement of effective safeguards for nuclear programs administered by the Agency or under its purview. He took great pains to assure the delegates that the "more basic purpose [of the IAEA] is the positive and creative development of the atomic era for human prosperity and welfare… there is much to be done, much to be learned before the atom can be widely and economically used for power. It is the duty of the Agency to hasten the doing and to hasten the learning."[17] This was translated into the Agency's charter as a dual mission of promoting nuclear energy and safeguarding peaceful use, creating internal tensions in the Agency that have never been satisfactorily resolved and which affected the negotiations over the safeguards system.

For example, as the IAEA safeguards negotiations progressed, major questions concerning safeguards were resolved in a way to provide the least hindrance to national nuclear development. When the notion of safeguarding source materials was proposed, objections by France and India were enough to kill the idea. Likewise, the no-

---

16. Seymour Hersh, *The Samson Option*, New York: Random House, 1991, p. 227.

17. "Opening of Discussions on Statute of International Atomic Energy Agency," *Department of State Bulletin*, Vol. 35, No. 902, October 8, 1956, p. 540, available from *ia802307.us.archive.org/10/items/departmentofstat3556unit/departmentofstat3556unit.pdf*.

tion of whether safeguards should apply to all states was defeated in favor of requiring safeguards only for countries receiving nuclear assistance from the IAEA. This meant that rich countries with highly developed nuclear programs could keep inspectors away from them while requiring such inspections for lesser developed countries. This discriminatory practice was carried over in principle into the Treaty on the Non-Proliferation of Nuclear Weapons (NPT), which divided parties to the treaty into weapons states and non-weapons states with different safeguards rules for each. To this practice, Homi Sethna, the head of India's nuclear program, issued a prescient comment: "This division into haves and have-nots will create lasting tensions that would only get worse with time."[18]

In dealing with the issue of stockpiling fissionable materials, Ambassador Wadsworth produced an agreement (contained in Article XII of the IAEA Charter) that enables the Agency "to require the deposit with the Agency of any excess of any fissionable materials recovered or produced as a by-product over what is needed [for peaceful purposes]…in order to prevent stockpiling of these materials." But the Agency has never produced a storage regime, so this article of the charter has never been implemented. Thus, the Agency has allowed Japan, for instance, to accumulate more than 9 tons of plutonium at home (and an additional 38 tons stored in England and France that will ultimately be returned to Japan) from which thousands of nuclear weapons could be manufactured.[19] A recent proposal to deal with this problem would use a bookkeeping scheme to give the IAEA virtual custody of that part of Japan's plutonium that would not be specifically identified by Japan as needed

---

18. George Perkovich, *India's Nuclear Bomb*, Berkeley: University of California Press, 1999, p.29.

19. Henry Sokolski, "A Plutonium Rich Asia," *National Review Online*, September 24, 2014, available from *http://www.npolicy.org/article.php?aid=1261&rt=&key=a%20plutonium%20rich%20asia&sec=article&author=*.

for future peaceful use.[20] The Agency would give up parts of its custody as needed by Japan for specific peaceful use. The devil is in the details, and whether Japan would seriously consider this virtual barrier to full and immediate access to its materials is unclear. But it begs the question of why the Agency allowed such stockpiles to be created in the first place in the absence of the full implementation of Article XII of the charter.

The charter of the IAEA that was formally adopted in 1957, as mentioned earlier, gave the Agency the dual missions of promoting nuclear energy and safeguarding peaceful use. The earlier warnings that promotion of nuclear energy would spread bomb-making technology were buried under a drive for future power and the prospect of future profits. The insidiousness of the dual missions became explicit when, in 2005, the then-head of the Agency, Mohammed El Baradei—whose relationship with the George W. Bush administration had up to then been problematic—publicly endorsed the nuclear deal between the United States and India,[21] despite previous UN resolutions passed in the wake of India's 1998 nuclear tests that called on India to adopt full scope safeguards and sign the NPT. One could be forgiven for concluding that the support of the IAEA for nonproliferation norms was conditional, depending on whose nuclear ox was being gored and who stood to gain from the expansion of nuclear power.

---

20. Fred McGoldrick, "IAEA Custody of Japanese Plutonium Stocks: Strengthening Confidence and Transparency," *Arms Control Today*, Vol. 33, No. 1, September 2014, available from *legacy.armscontrol.org/act/2014_09/Features/IAEA-Custody-of-Japanese-Plutonium-Stocks_Strengthening-Confidence-and-Transparency*.

21. "India-U.S. Nuclear Deal a Step Forward: El Baradei," *The Hindu* (Chennai), June 15, 2006, available from *www.thehindu.com/todays-paper/tp-international/indiaus-nuclear-deal-a-step-forward-el-baradei/article3119546.ece*.

The NPT

By the early 1960s, it was apparent that more countries would make nuclear weapons besides the United States, USSR, United Kingdom (which exploded its first bomb in 1952), and France (whose first test came in 1960). It was understood that support for nuclear disarmament among the states that had already made weapons was intimately tied to the prevention of weapons proliferation. In 1961, the UN General Assembly approved a resolution introduced by the Irish delegation calling on all states to refrain from the transfer or acquisition of nuclear weapons. This triggered more than 6 years of negotiations—led by the United States, USSR, and United Kingdom—that resulted in the NPT, which was opened for signature on July 1, 1968.[22] On the first day, 62 countries signed on and the treaty ultimately became the core of an international regime for fostering nuclear nonproliferation.

The treaty's aim, like that of the IAEA charter, was to prevent or restrain nuclear weapons proliferation without impeding nuclear development and thus required further development of an effective safeguards regime for NPT parties. A general safeguards objective was stated as follows for non-weapons state parties to the NPT: "The timely detection of the diversion of significant quantities of nuclear materials from peaceful activities...and deterrence of such diversion by the risk of early detection."[23]

In order for the IAEA to develop safeguards standards following the introduction of its safeguards system pursuant to the NPT, it was necessary to establish detection goals. Accordingly, in 1975, the Agency established the Standing Advisory Group on Safeguards Implementation (SAGSI), a group of safeguards experts from

---

22. Federation of American Scientists, "Treaty on the Non-Proliferation of Nuclear Weapons (NPT)," available from *www.fas.org/nuke/control/npt/*.

23. "The Technical Objective of Safeguards," *IAEA Bulletin*, Vol. 17, No. 2, April 1975, pp. 13-17, available from *www.iaea.org/sites/default/files/17203401317.pdf*.

IAEA member states, appointed by the Director General, to advise on safeguards issues.[24] SAGSI developed the notion of a Significant Quantity (SQ) of fissile materials (the approximate amount of material needed for a nuclear explosion), established timeliness goals, and laid out the format for reporting on safeguards performance in the general IAEA Safeguards Implementation Report.

Safeguards Standards[25]

In 1977, the formal detection goal of the Agency's safeguards system was stated as a 90-95% probability of detecting a diversion of one SQ of Special Nuclear Material (SNM), with a false alarm probability of less than 5%, and with a detection time less than or equal to the time needed to convert the SNM to a nuclear explosive device. The definition of one SQ of plutonium containing less than 80% plutonium-238 was set at 8 kilograms; the definition of one SQ of HEU was set at 25 kilograms of U-235. The conversion times were set at 7-10 days for metal; 1-3 days for oxides or nitrates; and 1-3 months for irradiated fuel.

There are a number of problems with these goals and their implementation: First, as to SQs, it is well known that nuclear weapons states have constructed working weapons that use much less than one SQ. Second, in a plant processing large amounts of material, the minimum detectable diversion will be much larger than one SQ. Third, the timeliness goals are unrealistic because they depend on the taking and analysis of material balances, which are normally

---

24. John Carlson, "SAGSI: Its Role and Contribution to Safeguards Development," Barton, Australia: Australian Safeguards and Nonproliferation Office, 2007, available from *www.dfat.gov.au/international-relations/security/asno/Documents/SAGSI_role_contribution_safeguards_dev.pdf*.

25. International Atomic Energy Agency, "The Present Status of IAEA Safeguards on Nuclear Fuel Cycle Facilities," *IAEA Bulletin*, Vol. 22, No. 3/4, August 1980, pp. 2-40, available from *www.iaea.org/sites/default/files/publications/magazines/bulletin/bull22-3/223_403400240.pdf*.

done annually while a diversion can occur at any time. One can overcome this problem by increasing the number of inventory takings, but this is expensive and is resisted by plant operators. Finally, in practice the technical requirements or goals are almost never met, partly because safeguards agreements are negotiated, as is the facility attachment laying out the actual inspection schedule. Thus, the argument over discrimination by the nuclear haves against the nuclear have-nots is played out in the negotiations over safeguards implementation, and plant operators (backed up by their governments) tend to resist effective safeguards agreements as intrusions into their domain. Indeed, the question of sovereignty hangs over the entire safeguards system.

Thus, the notion of inspectors roaming at will over the territory of a nuclear state to try to find clandestine materials or facilities has been resisted from the beginning, and safeguards were understood to be limited to declared materials and facilities. While some improvement has occurred thanks to the voluntary Additional Protocol introduced by the IAEA following the post-Gulf War revelations about Iraq's clandestine nuclear program, countries of high proliferation concern have not signed and ratified the Protocol. The Agency does have the right under its charter to undertake a special inspection to resolve safeguards ambiguities, but usually must first gain the cooperation of the inspected state. States also have the right to object to particular inspectors designated for it by the IAEA. Moreover, safeguards are not permanent under the NPT, so if a country leaves the treaty it is not obligated to continue inspections.

It should also be noted that NPT safeguards follow the nuclear materials in a country, so if a facility has no declared nuclear material in it, the facility is not subject to inspection unless the IAEA has reason to believe that materials are or have been present and the state involved agrees to the inspection. If there is a disagreement, the IAEA can go to the UN Security Council and attempt to obtain a resolution putting the state under threat of sanctions if it does not comply.

The bottom line on safeguards: They are useful and mandatory for providing some level of confidence that a peaceful nuclear program of one's neighbor is not a cover for a clandestine weapons program, but no country is likely to rely on it exclusively for that purpose. That is why restrictions on nuclear trade and development are important elements of a nonproliferation regime worthy of the name.

The Beginning of the NPT

The early history of the IAEA and the safeguards regime has led to weaknesses in protecting the world from the proliferation of nuclear weapons. Since the history of safeguards is intimately intertwined with the history of the NPT, it should not come as a surprise that the design of the NPT itself has occasionally been a source of grief for the nonproliferation that it was supposed to prevent.

Among the first steps toward the treaty was a unanimously adopted 1961 UN General Assembly resolution introduced by the Irish delegation that called on all states to conclude an international agreement to refrain from transferring or acquiring nuclear weapons.[26] In negotiating the agreement, the United States wanted language that would have allowed it to create a multilateral nuclear force (MNF) stationed in Germany.[27] The Soviets, however, made it clear that they would oppose creating any treaty that would allow the possibility of Germans being in command of nuclear weapons, but they could tolerate U.S. nuclear weapons in Europe, controlled and commanded by the Americans. Accordingly, the United States gave up on the idea of an MNF as long as the Soviets would cooperate on negotiating the NPT.[28]

---

26. Arms Control Association, "Timeline of the NPT," March 5, 2010, available from *www.armscontrol.org/system/files/NPT_Timeline.pdf.*

27. John Barton and Lawrence Weiler, eds., *International Arms Control: Issues and Agreements*, Stanford, CA: Stanford University Press, 1976, p. 296.

28. Ibid, p. 297.

The negotiations began within the (then 11 nation) UN Conference on Disarmament meeting in Italy in July 1965. The non-nuclear nations demanded from the beginning that they would support non-proliferation only if it was coupled to progress in disarmament by the nuclear nations. Specifically, the eight non-aligned countries in the Conference demanded "tangible steps to halt the nuclear arms race and to limit, reduce, and eliminate stockpiles of nuclear weapons and their means of delivery."[29]

Thus, the NPT would have to contain a grand bargain in which the non-weapons states pledge to give up their right to acquire nuclear weapons and the weapons states give up their right to retain them. The problems are, as usual, in the details.

Problems with the NPT

The treaty suffers from the same problem as the charter of the IAEA. It looks with approval on, and is intended to foster, the spread of nuclear development—including technologies that have both peaceful as well as weapons applications—as long as such development is "effectively" safeguarded. The preamble to the treaty affirms

> the principle that the benefits of peaceful applications of nuclear technology, including any technological by-products which may be derived by nuclear weapon States from the development of nuclear explosive devices, should be available for peaceful purposes to all Parties of the Treaty, whether nuclear-weapon or non-nuclear weapon States [and that]...all parties to the Treaty are entitled to participate in the fullest possible exchange of scientific

---

29. United Nations Conference of the Eighteen-Nation Committee on Disarmament, Final Verbatim Record of the Three Hundred and Thirtieth Meeting, ENDC/PV.330, September 14, 1967, p. 6, available from *quod.lib.umich.edu/e/endc/4918260.0330.001?rgn=main;view=fulltext*.

information for, and to contribute alone or in cooperation with other States to, the further development of the applications of atomic energy for peaceful purposes.[30]

This open door to advanced nuclear development, which can be interpreted in accordance with Article III of the treaty as including safeguarded enrichment and reprocessing technology, is essentially locked open by Article IV of the treaty, which establishes "the inalienable right of all the Parties to the Treaty to develop research, production, and use of nuclear energy for peaceful purposes, without discrimination and in conformity with Articles I and II of the Treaty" (that is, for peaceful purposes and under safeguards).[31]

Article IV, Paragraph 2 underscores the Preamble by giving all parties to the treaty the right to participate in the "fullest possible exchange of equipment, materials, and scientific and technological information for the peaceful uses of atomic energy"; and puts all parties in a position to do so under obligation to cooperate in the "further development of the applications of nuclear energy for peaceful purposes, especially in the territories of non-nuclear weapon States Party to the Treaty..."[32] It is this part of the treaty that is currently being cited by Iran in support of their claim that they have an inalienable right to enrich uranium for peaceful purposes, although a counter argument can be made that the Iranian centrifuge program is part of a weapons program and is therefore not legitimate under the NPT. This argument is likely to be settled ultimately by the jurisprudential adage that "possession is nine-tenths of the law," and no one appears to be in a position to take an unwilling Iran's centrifuges away from it by any means short of war.

---

30. *Treaty on the Non-Proliferation of Nuclear Weapons*, July 1, 1968, entered into force on March 5, 1970, 21 U.S.T. 483, 729 UNT.S. 161.

31. Ibid.

32. Ibid.

On the other hand, Iran has suffered significant economic damage as a result of sanctions imposed on it for being in technical violation of its safeguard obligations (in not having informed the IAEA of the existence of its enrichment program for about 15 years prior to 2003) and for ignoring UN Security Council resolutions demanding that Iran suspend its enrichment operation.[33] This economic pressure, among other things, induced Iran to enter into a years-long negotiation with six other countries (the P5+1) that ultimately resulted in the Joint Comprehensive Plan of Action (JCPOA) signed in 2015 and endorsed by all the negotiating parties plus the European Union. (See *www.state.gov/documents/organization/245317.pdf*)

The JCPOA will significantly reduce Iran's nuclear materials and production capabilities for at least the next ten years and provides for broad monitoring by the IAEA of Iran's compliance with the agreement. What will happen after the expiration of the agreement is unclear, but the need to have undergone this special effort to prevent further proliferation in the short term in the Middle East underscores the failure of the NPT and the IAEA to establish a tighter safeguards system in advance of the decisions to allow the sale of nuclear technology all over the world.

Even the fear of Improvised Nuclear Explosive Devices (INEDs), which would not exist were it not for the production of weapons-usable materials that are unneeded for peaceful applications, has been insufficient motivation for NPT parties to commit to stop such production. Instead, more controls (e.g., under UN Security Council Resolution 1540) are prescribed that depend on the same spotty implementation mechanisms that have been ineffective in preventing past proliferation. It is unclear what would enable nuclear states to stop producing more weapons-usable materials or to stop proliferating advanced nuclear technology that carries with it the threat of a weapons capability if not the weapons themselves. In that re-

---

33. Arms Control Association, "UN Security Council Resolutions on Iran," August 6, 2012, available from *www.armscontrol.org/factsheets/Security-Council-Resolutions-on-Iran.*

spect, the treaty's role in the promotion of nuclear energy, like that of the IAEA, can act as a spur to, as well as a blocker of, nuclear weapons proliferation.

The treaty also suffers from a problem inherent in many treaties: The failure to prevent a party from leaving the treaty without penalty (Article X simply requires three months notice and a statement suggesting a reason having to do with a possible nuclear threat). The lack of a penalty for leaving the NPT removes one disincentive for a non-weapons state using its treaty status to receive nuclear technology assistance with the intent of then leaving the treaty so it can exploit its new knowledge for weaponry purposes. It is the case that safeguard violations may be referred to the UN Security Council (UNSC) for action, as was done in Iran's case, but the Council cannot effectively take action if the violator is an ally of a permanent member of the UNSC who, by such membership, possesses veto power. Thus, violations are ultimately left to coalitions of the willing, which would have been the enforcement mechanism to prevent proliferation if the NPT did not exist. In the end, therefore, it is alliances that matter, although the treaty provides a useful organizational principle and creation of an international norm around which the power of an alliance can be exercised. Unfortunately, and perhaps unavoidably, if one of the permanent members of the UNSC violates the treaty, power politics is the likely tool for mitigation, and so it is no surprise that nuclear weapons states continue to possess their nuclear arsenals nearly half a century after the treaty went into force.

Indeed, even though the NPT Article VI requires good faith negotiations among the weapons states toward eliminating their nuclear arsenals, progress in this direction has had little connection to the existence of Article VI and is more the result of changed political conditions among and within the weapons state parties themselves. Nonetheless, as an expression of world aspiration, Article VI remains a useful reminder of the reasons why the NPT came into existence in the first place. But no one should consider it a substitute for the difficult agreements that are required for long-term peace and safety. Such difficulty combined with the desire for profits from nuclear and other trade has enabled the countries that have eschewed becoming committed parties to the NPT (India, Pakistan, Israel, and North Korea) to make nuclear weapons and prosper via economic relations with NPT parties. Money and goods are fungible. An NPT party that provides trade and assistance to a non-NPT party allows the latter to devote resources to its nuclear weapons program. This is not considered a violation of an NPT obligation under Article I or II. Much of the cynicism and hypocrisy that surrounds the NPT is traceable to the built-in discrimination and inequality in both the architecture of the treaty and the manner in which the treaty's provisions are interpreted and adjudicated.

Many of the issues raised by the treaty and the safeguards system were foreseen by some of the scientists who worked on the Manhattan Project. Their conclusion that proliferation was inevitable if national nuclear programs were allowed to exist and develop was prescient. The genie is indeed out of the bottle, and in a world where a majority of the American public would condone the use of nuclear weapons in war when they perceive a clear benefit over the use of conventional weapons,[34] putting the genie back is a daunting if not impossible task, even in circumstances where some former leaders who once argued for the construction and use of such weapons have belatedly changed their minds. But to paraphrase a much quoted statement about armies and war, you fight proliferation with the nonproliferation regime you have, not with the regime you want. Until the world becomes as sensitized to the risks of nuclear commerce and expansion as it is to the threat of nuclear war, especially at a time when global warming has resulted in an evident need to reduce dependence on fossil fueled energy, a world free of nuclear weapons will continue to be little more than an aspiration.

---

34. Darryl Press, Scott Sagan, and Benjamin Valentino: "Atomic Aversion: Experimental Evidence on Taboos, Traditions, and the Non-Use of Nuclear Weapons," *American Political Science Review*, Vol 107, No. 1, February 2013, pp. 108-206, available from *https://cisac.fsi.stanford.edu/sites/default/files/FINAL_APSR_Atomic_Aversion.pdf*.

CHAPTER 4

NPT'S Naval Nuclear Propulsion Loophole

Jeffrey M. Kaplow

Nuclear-powered submarines have long been exclusively the province of the established nuclear weapons states. But this small club is poised to expand. India is finally conducting sea trials of its long-delayed indigenous nuclear submarine, Brazil recently opened the shipyard that it hopes will construct five nuclear submarines over the next ten years, and Argentina and Iran also have expressed interest in deploying nuclear subs in the future.[1]

Beyond the concern this raises about the possibility of a new naval arms race, a number of analysts have pointed to the potential proliferation risk associated with nuclear submarines.[2] The Nuclear Non-

---

1. Rajat Pandit, "India's First Indigenous Nuclear Submarine Gears up for Maiden Sea Trials," *Times of India,* December 15, 2014, available from *timesofindia.indiatimes.com/india/Indias-first-indigenous-nuclear-submarine-gears-up-for-maiden-sea-trials/articleshow/45517702.cms*; Robin Yapp, "Argentina Developing Nuclear-Powered Submarine," *Telegraph* (London), August 2, 2011, available from *www.telegraph.co.uk/news/worldnews/southamerica/argentina/8677600/Argentina-developing-nuclear-powered-submarine.html*; Olli Heinonen, "Nuclear Submarine Program Surfaces in Iran," *Power & Policy* (blog), July 23, 2012, available from *http://belfercenter.hks.harvard.edu/publication/22207/nuclear_submarine_program_surfaces_in_iran.html*; and "Brazil's Rousseff Inaugurates Nuclear Sub Shipyard," *AFP*, December 12, 2014, available from *news.yahoo.com/brazils-rousseff-inaugurates-nuclear-sub-shipyard-224338823.html*.

2. See, for example, James Clay Moltz, "Closing the NPT Loophole on Exports of Naval Propulsion Reactors," *Nonproliferation Review,* Vol. 6, No. 1, Fall

proliferation Treaty (NPT) allows states to exempt nuclear material from international safeguards for use in nuclear submarines. The nuclear fuel that powers naval reactors is also useful in weapons work, and the nonproliferation community has long worried that exempted material could be diverted to a nuclear weapons program without the knowledge of inspectors. Naval nuclear propulsion, then, may represent a dangerous loophole in the NPT.[3]

The naval nuclear propulsion loophole, however, could function as a kind of canary in the coal mine: Any attempt by a proliferant state to take nuclear material out of safeguards for a nuclear submarine program—at least in the present international security environment—may be seen as a significant step toward the development of nuclear weapons. This feature of the naval propulsion loophole makes it a less desirable pathway to a weapon for potential proliferants. States with nuclear weapons aspirations probably would prefer an approach that would not be as quickly discovered, such as the use of covert facilities or the acquisition of sensitive nuclear materials from other states.

The alerting power of the naval nuclear propulsion loophole today, however, is partly a function of its novelty. No state has yet taken advantage of the ability to exempt material for use in military naval reactors, and only the P-5 nuclear weapons states currently deploy nuclear submarines. If the exercise of the naval propulsion exemp-

---

1998, pp. 108–114; Chunyan Ma and Frank von Hippel, "Ending the Production of Highly Enriched Uranium for Naval Reactors," *Nonproliferation Review*, Vol. 8, No. 1, Spring 2001, pp. 86–101; Greg Thielmann and Serena Kelleher-Vergantini, "The Naval Nuclear Reactor Threat to the NPT," *Threat Assessment Brief*, Arms Control Association, July 24, 2013, available from www.armscontrol.org/files/TAB_Naval_Nuclear_Reactor_Threat_to_the_NPT_2013.pdf; and John M. Lamb, "Roiling the Arms Control Waters," *Bulletin of the Atomic Scientists*, Vol. 43, No. 8, October 1987, pp. 17–19.

3. Naval nuclear propulsion is the use of a nuclear reactor to power a naval vessel. While most naval reactors power submarines, both the United States and Russia have produced nuclear-powered surface ships, and Russia still operates a fleet of nuclear icebreakers.

tion comes to be seen as acceptable or normal, or if more states begin deploying nuclear submarines, then the loophole could become much more dangerous. An Iranian exemption of nuclear material for a supposed submarine effort, for example, would set off fewer alarm bells if its rivals also were pursuing nuclear-powered subs. And so the international community is right to attempt to dissuade states from removing nuclear material from safeguards for naval propulsion, and to pursue other avenues for closing or limiting the loophole.

This chapter proceeds in three parts. First, I discuss the origins of the naval nuclear propulsion loophole in more detail and point to several ways in which it differs from other NPT loopholes. Next, I survey the handful of states that have expressed interest in nuclear submarines, highlighting the nonproliferation implications of their naval propulsion programs. Finally, I describe several policy options for narrowing the loophole or for closing it altogether.

The Origins and Consequences of the Naval Nuclear Propulsion Loophole

The naval nuclear propulsion loophole differs from other gaps in the nuclear nonproliferation regime in several ways, with important implications for how the loophole is perceived by the international community and in how, ultimately, it can be filled. First, the safeguards exemption for naval reactors is a sin of omission—it is not made explicit in the NPT. This means that the exemption is actually somewhat broader than is commonly realized, and also that debate about the loophole cannot be resolved by reference to treaty text. Second, this gap in the treaty was no accident; it was quite explicitly designed into the NPT to smooth the path for state ratification and to respond to the objections of allies. Third, the potential danger of the loophole was recognized at the time the agreement was drafted. The United States and others calculated that the benefit to nuclear nonproliferation goals in winning the NPT adherence of key states outweighed the future risk that the loophole would be used to evade international safeguards. So far, at least, this calculation seems to have been correct. Finally, the declaration requirements associated with this loophole make it highly alerting.

A Sin of Omission

There are two broad categories of loopholes in international agreements. Perhaps the most common type of loophole is invoked by the text of the treaty itself, or comes about as a result of disagreements about the correct interpretation of treaty language. The withdrawal clause of the NPT, laid out in Article X, is a loophole of this type. Similarly, some argue that the nuclear weapons states' reluctance to disarm is a kind of loophole resulting from a particular interpretation of their requirements under Article VI.[4]

Another type of loophole is created when the text of an international agreement fails to explicitly address some possible state action. Loopholes formed in this way do not necessarily imply that the drafters of the treaty did not consider the issue—such loopholes may be intentional or not. No treaty is exhaustive, and states must make decisions about what issues to cover explicitly in the text of an agreement.

The safeguards exemption for naval nuclear propulsion is of this latter type: It is a sin of omission. The NPT simply does not address the military uses of nuclear technology beyond nuclear weapons. It was left, then, to the International Atomic Energy Agency (IAEA) to create rules about how to safeguard enriched uranium intended for use in military naval reactors. Recognizing that international

---

4. See, for example, David A. Koplow, "Parsing Good Faith: Has the United States Violated Article VI of the Nuclear Non-Proliferation Treaty?" *Wisconsin Law Review,* Vol. 301, March/April 1993, pp. 301–394. Of course, others disagree. See Christopher A. Ford, "Debating Disarmament: Interpreting Article VI of the Treaty on the Non-Proliferation of Nuclear Weapons," *Nonproliferation Review,* Vol. 14, No. 3, November 2007, pp. 401–428.

inspections of military facilities would be a non-starter, the IAEA relies instead on state declarations. When exempting nuclear materials from safeguards for non-explosive military use, states must declare the activity and the amount of material employed, provide assurances that the material will not be used for nuclear weapons, and agree to reinstate safeguards on the material when its use for military purposes concludes.[5] The IAEA, however, does not attempt to verify these declarations, and so states may see this exemption as a convenient way to divert nuclear material for use in a covert weapons program.

The distinction between gaps in a treaty and more explicit loopholes is important for three reasons. First, when loopholes are not addressed in an international agreement, no amount of legal wrangling over the treaty text will settle the issue. Justification for explicit loopholes often comes down to a debate over the original intent of the treaty language, or over the broader context of particular treaty clauses.[6] This kind of argument is largely avoided when loopholes are simply not covered by the treaty. When non-nuclear weapons states within the NPT have announced their interest in nuclear submarines, for example, the public debate has centered on the nonproliferation or other consequences of that behavior, rather than its legality. That is, taking advantage of the naval propulsion exemption may be unwise, but it is not illegal.

Second, there may be a broader consensus about the existence of a loophole when it is not addressed by the treaty at all. In the case of the naval nuclear propulsion loophole, however, the broad agree-

---

5. International Atomic Energy Agency, *INFCIRC/153 (Corrected): The Structure and Content of Agreements Between the Agency and States Required in Connection with the Treaty on the Non-Proliferation of Nuclear Weapons*, Vienna: International Atomic Energy Agency, 1972.

6. The Vienna Convention on the Law of Treaties allows the negotiating history of an international agreement to clarify state obligations. Arguments relying on the intent of treaty drafters are routinely made outside of the courts, as well. See, for example, Ford, "Debating Disarmament."

ment that a loophole exists has not yet led to a concerted attempt at filling the gap in the treaty. This lack of action—in the face of many proposals—may stem from the perception that the loophole does not at this time constitute a significant threat to nonproliferation goals.

Finally, loopholes created by omission are often much broader than those resulting from explicit treaty text. With no treaty language to constrain them, states can take advantage of entire issue areas in which to act without fear of legal transgression. Because it is created by a gap in the treaty, the naval nuclear propulsion loophole is in fact quite a bit broader than it first appears. In contemporary policy debates, the NPT's failure to address non-explosive military uses of nuclear technology is most relevant to naval nuclear propulsion, but there are potentially a number of other applications that qualify for the exemption. Material to power nuclear reactors for military spacecraft might be exempted: Both the United States and Soviet Union had long-running space nuclear propulsion programs, and other countries, such as China and France, investigated the technology.[7] The exemption also would apply to material destined for military reactors intended for radiation testing or to power a military base. The United States, for example, ran a long-standing program to develop small nuclear power reactors for military installations.[8] While the United States was aware of these broader uses for the military exemption, it preferred to keep the focus on naval reactors. For example, a now-declassified State Department cable cautioned the U.S. Embassy in Tokyo in 1976 that the exemption had only

---

7. On space nuclear propulsion efforts in the civilian sphere, see A. Stanculescu, et al., *The Role of Nuclear Power and Nuclear Propulsion in the Peaceful Exploration of Space*, Vienna: International Atomic Energy Agency, 2005.

8. See Lawrence H. Suid, *The Army's Nuclear Power Program: The Evolution of a Support Agency*, Westport, CT: Greenwood Press, 1990. Although the Army Nuclear Power Program concluded in the late 1970s, analysts continue to discuss the possibility of powering military bases with nuclear reactors. See Marcus King, LaVar Huntzinger, and Thoi Nguyen, *Feasibility of Nuclear Power on U.S. Military Installations*, Arlington, VA: CNA Corporation, March 2011.

been publicly linked to naval nuclear propulsion. It instructed that other possible applications should not be volunteered to Japanese government officials, but could be acknowledged if asked.[9]

A Loophole by Design

The NPT's failure to address military non-explosive uses of nuclear technology was not an accident. Early drafts of the treaty included language that would have required non-nuclear weapons states to put all of their nuclear activities under safeguards, eliminating the possibility of exempting nuclear material from safeguards for any reason.[10] By the time the NPT opened for signature in 1968, Article III limited safeguards for non-nuclear weapons states to "all source or special fissionable material in all *peaceful* nuclear activities," (emphasis added) thus excluding military non-explosive uses such as naval propulsion.[11]

Naval nuclear propulsion ultimately was left out of the NPT because of the complex dynamics of multilateral treaty negotiations. Two factors in particular influenced the decision to allow the military exemption. First, the United States recognized that the NPT would only be effective to the extent that it received widespread international adherence, and so Washington was focused on win-

---

9. U.S. Department of State, "State Department Cable 040620 to Embassy Tokyo," February 20, 1976, Confidential, Central Foreign Policy Files, Record Group 59, National Archives at College Park, MD, Retrieved from the Access to Archival Databases, available from *www.archives.gov*.

10. U.S. Department of State, Draft Position Paper, "Non-Proliferation of Nuclear Weapons," August 14, 1964, Secret Noforn, Document 44, *Foreign Relations of the United States, 1964-1968*, Volume XI, Arms Control and Disarmament, available from *history.state.gov/historicaldocuments/frus1964-68v11/d44* and U.S. Arms Control and Disarmament Agency, Draft Position Paper, "Safeguards on Peaceful Nuclear Activities," December 22, 1965, Confidential.

11. *Treaty on the Non-Proliferation of Nuclear Weapons*, July 1, 1968, entered into force on March 5, 1970, 21 U.S.T. 483, 729 UNT.S. 161.

ning the approval of key allies for the proposed treaty. Italy and the Netherlands, however, hoped to pursue nuclear-powered naval vessels in the future, while the United Kingdom worried that treaty language would complicate the import of naval reactors from the United States.[12] By leaving a gap in the treaty, the NPT's drafters helped to allay the concerns of these important allies.[13] Even the requirement that nonweapons states place their *peaceful* nuclear activities under IAEA safeguards was something of a victory; U.S. allies had insisted on removing language calling for mandatory safeguards from early drafts of the NPT.[14]

Second, the NPT drew criticism from some quarters for putting in place a two-tiered system, in which the five recognized nuclear

---

12. Moltz, p. 109; David Fischer, *History of the International Atomic Energy Agency: The First Forty Years*, Vienna: International Atomic Energy Agency, 1997, p. 272, available from *www-pub.iaea.org/mtcd/publications/pdf/pub1032_web.pdf*; and U.S. Arms Control and Disarmament Agency, Draft Position Paper, "Safeguards on Peaceful Nuclear Activities." A potential concern might have been the effect of NPT safeguards provisions on plans for a NATO multilateral force—this was a major issue in the NPT negotiations as a whole—but by this time the possibility of multilateral nuclear submarines had been taken off the table. See Wilfrid L. Kohl, "Nuclear Sharing in Nato and the Multilateral Force," *Political Science Quarterly*, Vol. 80, No. 1, March 1965, pp. 94–95.

13. Of course, these states may ultimately have signed the treaty even without the presence of the naval reactor exemption; certainly some had multiple concerns about the NPT even in its final form. Italy, for example, "had been quite difficult" in negotiating the treaty, according to the Director of the U.S. Arms Control and Disarmament Agency. See "Memorandum for the Record of the 548th Meeting of the National Security Council," March 27, 1968, Secret, Document 229, *Foreign Relations of the United States, 1964-1968*, Volume XI, Arms Control and Disarmament, available from *history.state.gov/historicaldocuments/frus1964-68v11/d229*. On Italy's stance toward NPT ratification more broadly, see U.S. Department of State, Bureau of Intelligence and Research, "Intelligence Note 605, Italian Parliament Gives Overwhelming Backing to NPT," July 31, 1968, available from *www2.gwu.edu/~nsarchiv/nukevault/ebb253/doc31.pdf*.

14. "Editorial Note," Document 92, *Foreign Relations of the United States, 1964–1968*, Volume XI, Arms Control and Disarmament, available from *history.state.gov/historicaldocuments/frus1964-68v11/d92*.

weapons states would be treated differently than other members. To some extent, this was unavoidable—the fundamental goal of the NPT, after all, was to prevent new countries from joining the select club of nuclear weapons states. But the treaty also created obligations for non-nuclear states that the weapons states did not share; foremost among these was the requirement that states outside the P-5 place their nuclear facilities under international safeguards. The United States tried to cushion the blow by voluntarily offering to implement IAEA safeguards at its civilian facilities, but no nuclear weapons state—including the United States—was willing to go further and allow inspectors to have access to sensitive military installations.[15] In this context, requiring nonweapons states to place non-explosive military activities under safeguards, or prohibiting such activities altogether, might have been seen as one more way in which the nonweapons states were being asked to bear a larger share of the burden under the treaty.

Anticipating a Loophole

Loopholes in international agreements often take countries by surprise; either they are not anticipated by the drafters of the treaty, or changes in technology or circumstances create new opportunities for states to evade the intent of the agreement. The naval nucle-

---

15. Proposals that would have required weapons states to place all peaceful nuclear activities under safeguards—just as nonweapons states were required to do—were rejected by the Soviet Union. See, for example, U.S. Arms Control and Disarmament Agency, "Memorandum, Director Foster to Secretary of State Rusk," January 11, 1967, Secret Limdis, Document 172, *Foreign Relations of the United States, 1964-1968*, Volume XI, Arms Control and Disarmament, available from *history.state.gov/historicaldocuments/frus1964-68v11/d172*. The nuclear weapons states all ultimately implemented voluntary safeguards at civilian nuclear facilities. For background on these agreements, see Adolf von Baeckmann, "IAEA Safeguards in Nuclear-Weapon States: A Review of Objectives, Purposes, and Achievements," *IAEA Bulletin*, Vol. 30, No. 1, March 1988, available from *www.iaea.org/sites/default/files/publications/magazines/bulletin/bull30-1/30103552224.pdf*.

ar propulsion loophole is different. Not only did the international community intentionally avoid addressing the naval propulsion issue, the risk that states would take advantage of this gap in the treaty was well understood at the time the NPT was being negotiated. A draft U.S. position paper from 1965 was quite explicit: "The U.S. position…is that we do not wish to create a possible loophole whereby a non-nuclear state might claim the right to exempt important nuclear facilities from safeguards, and…perhaps raise suspicions that clandestine nuclear weapons work was being carried out in those facilities."[16] At the same time, however, a spate of nuclear submarine programs seemed quite a long way off in the late 1960s; only the nuclear weapons states seemed likely to deploy naval propulsion reactors for the foreseeable future. That the United States ultimately allowed the loophole into the treaty—knowing its potential consequences—suggests that its benefits were thought to exceed its costs.

A Canary in the Coal Mine

Before taking advantage of the safeguards exemption for naval nuclear propulsion, an NPT member state must make a detailed declaration to the IAEA. This has two important consequences for the loophole. First, it means that actually exercising the loophole can be highly alerting, acting like a canary in the coal mine for nuclear weapons intentions. When a state invokes the naval propulsion exemption, it puts other member states on notice and signals a greater risk of treaty non-compliance or treaty withdrawal in the future. Second, because exempting nuclear material from safeguards is so alerting, it becomes much more costly for states to take advantage of the loophole. Countries using the exemption invite an international response: They may face pressure or threats aimed at reversing their decision, and these consequences may extend well

---

16. U.S. Arms Control and Disarmament Agency, Draft Position Paper, "Safeguards on Peaceful Nuclear Activities."

outside the bounds of the NPT. States considering taking advantage of the naval propulsion exemption may anticipate an international response and choose other pathways to a bomb. The fact that the loophole is highly alerting, then, may ultimately make it less likely to be invoked in the first place.

This is the silver lining of the naval nuclear propulsion exemption: A country that takes advantage of the exemption is likely to have its nuclear activities subjected to increased scrutiny. Diverting material via the naval propulsion exemption is thus a dangerous way for potential proliferants to kick-start a nuclear weapons program. Other plausible nuclear weapons pathways for NPT member states—a fully covert enrichment program, for example, or acquisition of sensitive nuclear materials from abroad—are probably less alerting, and potentially give other states less time to respond with pressure, sanctions, or attack. Knowing this, proliferant states probably will not choose to divert material under the naval propulsion exemption, opting for less alerting pathways to a bomb, and reducing the danger ultimately posed by the loophole.

The withdrawal clause of the NPT is another loophole of this kind, because it requires a public declaration. Even more than the naval propulsion loophole, it is highly alerting. Any state looking to exit the NPT is basically announcing its intention to pursue nuclear weapons, making withdrawal an unattractive option for proliferants. State leaders seem to agree: While NPT members have engaged in 10 nuclear weapons programs since 1970, only North Korea has opted to withdraw from the treaty, and then only after making substantial progress toward a nuclear weapon.[17]

An important caveat, however, applies to these NPT loopholes:

---

17. This count of nuclear weapons programs is updated from Dong-Joon Jo and Erik Gartzke, "Determinants of Nuclear Weapons Proliferation," *Journal of Conflict Resolution*, Vol. 51, No. 1, February 1, 2007, pp. 167–194. See also Jeffrey M. Kaplow, "State Compliance and the Track Record of the Nuclear Nonproliferation Regime," working paper of Ph. D. dissertation, University of California, San Diego, 2014, available from *dl.jkaplow.net/KaplowCh1.pdf*.

They are only alerting so long as they are rarely employed. If several states had recently left the NPT, for example to protest the lack of progress on nuclear disarmament, then the next withdrawal—even by a state with a latent nuclear capability—would be less alarming. Similarly, an Iranian decision today to exempt nuclear material from safeguards for a nuclear submarine effort would set off many alarm bells. The same decision would be a weaker indicator of nuclear weapons ambitions, however, if it followed similar exemptions by others in the region. The exercise of these loopholes serves to legitimize them, ultimately making them less informative about a state's intentions.

The naval nuclear propulsion exemption also becomes less alerting if a state can make a plausible case that a nuclear submarine program is militarily useful. Nuclear submarines have long represented an attractive military capability, particularly for states concerned with the survivability of a nuclear deterrent, because they can stay underwater for longer and venture further than their conventionally powered counterparts. The benefits of naval reactors, however, are more than matched by their significant cost in development and operation, and lower-cost air-independent propulsion technologies now represent a viable alternative to nuclear propulsion for most states.[18] Few non-nuclear states can make a reasonable claim that a nuclear submarine program is worth the cost, and this hurdle is likely to get even higher over time as alternative technologies become both cheaper and more effective.

A Survey of Nuclear Submarine Programs

Only five states—the United States, Russia, the United Kingdom, France, and China—currently deploy nuclear submarines with

---

18. Ma and von Hippel, pp. 97–98. On the difficulties that modern, non-nuclear attack submarines pose for U.S. anti-submarine warfare, see John R. Benedict, "The Unraveling and Revitalization of U.S. Navy Antisubmarine Warfare," *Naval War College Review*, Vol. 58, No. 2, Spring 2005, pp. 101–102.

their naval forces. Several other countries have expressed interest in naval nuclear propulsion over the years, however, with potential consequences for global nuclear nonproliferation efforts. In this section, I survey the nuclear submarine landscape beyond the P-5, with an emphasis on the nonproliferation implications of each state's naval nuclear propulsion efforts.

Canada

Canada does not currently have a nuclear submarine program, but it caught many by surprise when it announced in 1987 its intention to add 10–12 nuclear submarines to its naval forces, one piece of a larger strategy to reassert Canadian sovereignty over Arctic territories and improve Canada's deterrent posture.[19] Despite some opposition from Canada's allies, both France and the United Kingdom intended to compete to supply the submarines.[20] Canada's plans were ultimately abandoned—the shifting views of the Canadian public and the end of Cold War made additional military spending unpopular.

The Canada nuclear submarine episode served as a warning about the ease with which a dangerous precedent could be set, even by a state that was strongly supportive of global nonproliferation efforts. Canada would have been the first state to exempt nuclear material from safeguards for military use—the Canadian government even

---

19. Canada's Defense Minister in 2012 caused some controversy by implying Canadian interest in deploying nuclear submarines, but his comments were quickly walked back. See Laura Payton, "No Nuclear Sub Buy Planned, MacKay Affirms," *CBC News*, October 28, 2011, available from *www.cbc.ca/1.1043181*. On the 1987 defense strategy, see Canada Department of National Defence, *Challenge and Commitment: A Defence Policy for Canada*, Ottawa: Canadian Government Publishing Centre, 1987, pp. 52–53, available from *publications. gc.ca/collections/collection_2012/dn-nd/D2-73-1987-eng.pdf.*

20. Steve Shallhorn: "Standing up to the United States," *Bulletin of the Atomic Scientists,* Vol.43, No. 8, October 1987, pp. 16–17.

emphasized that it would set a positive example for other states considering exemptions—and this fact drew the attention and criticism of the nuclear nonproliferation community.[21] While Canada itself would not use the exemption to supply a covert nuclear weapons program, its foray into naval nuclear propulsion could have made it easier for others to do so. Removing nuclear material from IAEA safeguards potentially would legitimize the use of nuclear technology for military purposes within the NPT and provide an example for other, less trustworthy states to point to in justifying their actions.

In Canada's case, there was another precedent at play: The first sale of a nuclear submarine to a non-nuclear weapons state. Had the Canadian plan moved forward, it might have increased both the supply-and demand-side risk that additional states would acquire nuclear submarines. On the supply side, the nuclear weapons states had largely held firm in denying the sale of nuclear submarines or naval reactor technology to nonweapons states. This had always been something of an uneasy truce, however, because the nuclear weapons states have significant financial incentives to market such technology abroad. If this barrier were broken and nuclear submarine sales were to become seen as just another arms deal, there could be many additional states in line to acquire naval reactor technology. On the demand side, when new states acquire a nuclear submarine capability, their rivals may feel compelled to follow suit, both to nullify any potential strategic advantage and to satisfy domestic constituencies calling for military parity to be maintained. More broadly, the spread of naval nuclear propulsion may lead states to see nuclear submarines as a sign of international prestige or of status as a major power.

The spread of nuclear submarines indirectly increases the risk of

---

21. See, for example, Tariq Rauf and Marie-France Desjardins, "Canada's Nuclear Submarine Program: A New Proliferation Concern," *Arms Control Today*, Vol. 18, No. 10, December 1988, pp. 13–18; and William Epstein, "New Stance Tarnishes Canada's Reputation," *Bulletin of the Atomic Scientists*, Vol.43, No. 8, October 1987, pp. 11–12.

nuclear proliferation: More nonweapons states with nuclear submarines means more chances to divert nuclear material using the naval nuclear propulsion exemption. Perhaps more dangerous, however, is the ability of naval reactors to justify the enrichment of uranium at higher levels. Facilities producing fuel for power reactors generally output material enriched to no more than five percent uranium-235 (U-235), but most naval reactors call for at least 20 percent enriched uranium and some use weapons-grade material enriched to over 90 percent.[22] A state with a nuclear submarine may have a built-in excuse to, first, optimize uranium enrichment facilities for the production of material enriched to higher levels and, second, stockpile uranium enriched at those higher levels. The ability to produce higher enriched material dramatically increases a state's nuclear latency and its ability to quickly manufacture a nuclear weapon in a breakout scenario.

India

In 1988, India became the first state outside of the P-5 to operate a nuclear submarine, leasing a Charlie-class submarine from the Soviet Union. That boat was returned in 1991, but in 2012 India added a leased Russian Akula-class nuclear submarine to its fleet. Both deals offered a useful training platform and a source of technology transfer for India's indigenous nuclear submarine effort.[23] The first Indian-built nuclear submarine, the INS Arihant, was unveiled in 2009 and began sea trials in late 2014.[24]

---

22. Ma and von Hippel, p. 91.

23. Moltz, p. 110; and Yogesh Joshi, "Leased Sub Key to India's Naval Modernization," *World Politics Review*, June 1, 2012, available from *www.worldpoliticsreview.com/articles/12014/leased-sub-key-to-indias-naval-modernization*.

24. Pandit, "India's First Indigenous Nuclear Submarine Gears up for Maiden Sea Trials." The name *Arihant*, reassuringly, is Sanskrit for "destroyer of enemies."

Because India is outside the NPT and already possesses nuclear weapons, its development of nuclear submarines does not set a new precedent with respect to the military use of nuclear material under IAEA safeguards. India's nuclear propulsion ambitions do, however, have several indirect consequences for nuclear nonproliferation. First, India's efforts raise the profile of nuclear submarine technology and demonstrate to other states that this capability is attainable. This, in turn, may make nonweapons states within the NPT more likely to seek nuclear submarines of their own. Second, once India has a demonstrated nuclear submarine capability, it becomes a potential supplier of nuclear propulsion technology to other states, with all of the proliferation risks that implies. While onward proliferation is a concern, it is worth noting that India has shown restraint in other aspects of nuclear supply. Finally, Indian nuclear submarines potentially affect the strategic balance with Pakistan, and may prompt Islamabad to intensify its naval efforts. Should Pakistan seek its own nuclear submarine capability, it is likely to turn to China for assistance, further weakening the norm against the supply of naval nuclear propulsion technology and leading to an escalation of what is already a low-level naval arms race in the region.[25]

Brazil

Brazil's nuclear submarine effort dates from the late 1970s, part of the parallel nuclear program run by the nation's military services. The Brazilian navy's contribution to the program included both the development of gas centrifuge uranium enrichment technology and

---

25. Iskander Rehman, "Drowning Stability: The Perils of Naval Nuclearization and Brinkmanship in the Indian Ocean," *Naval War College Review*, Vol. 65, No. 4, Autumn 2012, pp. 64–88 and Yogesh Joshi and Frank O'Donnell, "India's Submarine Deterrent and Asian Nuclear Proliferation," *Survival*, Vol. 56, No. 4, August/September, 2014, pp. 157–174.

exploration of naval propulsion reactors.[26] The navy's nuclear submarine work persisted at a low level even as most of the military's nuclear efforts were shuttered in the late 1980s and early 1990s; the submarine program was revitalized in 2007 with the announcement of significant funding to build a prototype nuclear propulsion reactor. With new buy-in from the political leadership, progress in the nuclear submarine program has accelerated in recent years. A deal with France will provide assistance with the non-nuclear components of the submarine, and Brazil aims to complete the first of six planned subs by the mid-2020s.[27]

The rationale for a Brazilian nuclear submarine capability has never been completely clear, and still puzzles analysts. At a recent dialogue between Brazilian and U.S. defense experts and officials, the motivation for the submarine program led to some debate even within the Brazilian delegation. A Brazilian defense analyst "argued that Brazil not only lacked a compelling rationale for such submarines, but that the cost accounting for the program was nontransparent. ... The discussion revealed that there is little public knowledge of the Brazilian nuclear propulsion program, nor is there clarity on the full costs or rationale for the program."[28] Brazilian military and political leaders have spoken vaguely about Brazil's need to defend its maritime interests and national sovereignty, and more specifically about the defense of offshore oil and gas assets. Of course, the nuclear submarine effort long predates the discovery of those energy resources, and there may be better tools for the job than a naval capability that is generally seen as a way to project power

---

26. "Memorandum, Information for the President of Brazil, No. 011/85 from the National Security Council, Structure of the Parallel Nuclear Program," February 21, 1985, Wilson Center, History and Public Policy Program Digital Archive, available from *digitalarchive.wilsoncenter.org/document/116917*.

27. Thielmann and Kelleher-Vergantini, pp. 4-5.

28. Anne L. Clunan and Judith Tulkoff, *Perspectives on Global and Regional Security and Implications of Nuclear and Space Technologies*, Monterey, CA: Naval Postgraduate School, October 2014, p. 20, available from *calhoun.nps.edu/handle/10945/43786*.

far from a state's borders.[29] Fundamentally, the nuclear submarine program probably has more to do with Brazil's great power aspirations than with any of the stated military needs.[30]

Whatever its intended purpose, Brazil's long-delayed nuclear submarine is notable for its nonproliferation consequences: It is likely to be the first case in which nuclear material is exempted from international safeguards for military use. Like the Canadian flirtation with nuclear submarines described above, a move by Brazil to take advantage of the NPT's naval propulsion exemption would set a precedent for other non-nuclear weapons states, and so risk weakening IAEA safeguards generally. Unlike Canada, however, Brazil has not been a model nonproliferation citizen. It is widely considered to have had an active nuclear weapons program from the late-1970s through the 1980s. It was a late adherent to the NPT, finally joining the treaty in 1998, nearly the last nonweapons state to do so. Negotiations with the IAEA over safeguards procedures at Brazil's enrichment facility, typically managed without fanfare by mid-level officials, made international news because of Brazil's insistence that inspectors not be allowed to actually see the centrifuges operating at the plant.[31] And Brazil remains a high-profile holdout when it comes to the Additional Protocol to its IAEA safeguards agreement, a more stringent set of declaration and inspection requirements for nuclear activities that has been signed by 124 states.

---

29. Sarah Diehl and Eduardo Fujii, "Brazil's Pursuit of a Nuclear Submarine Raises Proliferation Concerns," *WMD Insights*, No. 23, March 2008, pp. 9–18.

30. For a full discussion of Brazil's motivations for seeking nuclear technology generally, including the nuclear submarine effort, see Togzhan Kassenova, *Brazil's Nuclear Kaleidoscope: An Evolving Identity*, Washington, DC: Carnegie Endowment for International Peace, 2014, available from c*http://carnegieendowment.org/2014/03/12/brazil-s-nuclear-kaleidoscope-evolving-identity-pub-54832.*

31. Sharon Squassoni and David Fite, "Brazil as Litmus Test: Resende and Restrictions on Uranium Enrichment," *Arms Control Today*, Vol. 35, No. 8, October 2005, available from *legacy.armscontrol.org/act/2005_10/Oct-Brazil.*

Given Brazil's controversial nuclear past, a decision to exempt nuclear material from safeguards for a submarine program would draw strong criticism from the nonproliferation community. But here, as in the case of Canada, the concern is less about the risk of Brazil itself diverting material to a nuclear weapons program, and more about the precedent this activity sets for other NPT member states. If Brazil takes advantage of this exemption, it makes it more likely that another state—such as Iran—follows suit, and at the same time makes the actions of that other state less alerting.[32] Once the transfer of material to a nuclear submarine program comes to be seen as acceptable, states may judge that they too can use the exemption without the international community assuming that they are working toward a nuclear weapon.

Iran

Iran is not known to have an active nuclear submarine program, but several high-level Iranian officials have expressed the intent to pursue naval nuclear propulsion, or at least to keep that option open.[33] Iran's interest in nuclear submarines has been met with a combination of alarm and skepticism by the international community. Alarm because an Iranian appeal to a nuclear submarine program seems to validate the worst fears of those concerned about the naval propulsion loophole. Iran, after all, is widely suspected to have harbored nuclear weapons ambitions and has been found to have violated its NPT commitments; the loophole would allow it to simply and legally remove nuclear material from international safeguards. It can take this step—and here the skepticism of the in-

---

32. The same potential consequences apply to the nuclear submarine ambitions of a state like Argentina. See Yapp, "Argentina Developing Nuclear-Powered Submarine."

33. Heinonen, "Nuclear Submarine Program Surfaces in Iran"; and "Iran Mulls Highly Enriched Uranium," *UPI*, April 17, 2013, available from *www.upi.com/Top_News/Special/2013/04/17/Iran-mulls-highly-enriched-uranium/82301366205307/*.

ternational community comes in—even if there is no actual nuclear submarine program.

And it gets worse: Even if nuclear material remains under safeguards, a supposed nuclear submarine program gives Iran an excuse to enrich uranium to higher levels, bringing it closer to the nuclear threshold if it should decide to push forward and build a weapon. Iranian officials already seem to be deploying the naval propulsion rationale for higher levels of enrichment. The Director of Iran's nuclear agency told reporters in 2013 that, "[a]t present, we have no enrichment plan for purity levels above 20 percent but when it comes to certain needs, for example, for some ships and submarines, if our researchers need to have a stronger underwater presence, we will have to make small engines which should be fueled by 45-56 percent enriched uranium."[34] Iran's parliament has done its part to support this position, approving a symbolic bill that would require the government to build and fuel nuclear-powered commercial naval vessels.[35] Nuclear material for this kind of non-military naval propulsion would not qualify for exemption from safeguards, but it still provides a useful justification for enrichment beyond 20 percent U-235.

Of course, these statements and parliamentary maneuvers occur in the context of Iran's ongoing standoff with the international community over its nuclear program, and so they may say more about Iran's negotiating strategy than any real-world plans for a nuclear submarine or for higher levels of uranium enrichment. Raising the specter of a nuclear submarine gives Iranian negotiators one more bargaining chip in service of a larger nuclear deal, and it is a bargaining chip that may cost almost nothing for Iran to give away if Iran does not actually intend to pursue a nuclear submarine or take advantage of the naval reactor exemption. This has been the domi-

---

34. Ibid.

35. Ali Akbar Dareini, "Iran Parliamentarians Call for Nuclear Ships," *Associated Press*, July 15, 2012, available from *news.yahoo.com/iran-parliamentarians-call-nuclear-ships-172547792.html*.

nant interpretation of Iranian hints about a nuclear submarine—a negotiating ploy and an excuse for enrichment, but nothing more.[36]

Actually exercising the naval propulsion loophole, then—exempting nuclear material and diverting it for weapons purposes—does not seem like the most likely scenario for Iran. But consider, anyway, what would happen if Iran today announced that it was exempting some enriched uranium from safeguards for the purposes of testing a naval nuclear reactor. Beyond the diplomatic protestations that would surely follow, the portion of the international community already skeptical of Iranian motives—the United States, Western Europe, and Israel most of all—would have to at least confront the strong possibility that Iran had taken the decision to pursue nuclear weapons. Talks would be suspended, sanctions would tighten, international tensions would increase, and the risk of military action to stave off Iranian nuclear weapons possession would jump dramatically. If Iran's goal were merely to remove enriched uranium from the watchful eye of the IAEA, then it would certainly succeed. But this small victory would come at a significant cost. While the invocation of a nuclear submarine program might provide enough of a fig leaf for Iran's actions that some states already sympathetic to Iran might defend its "right" to exempt material under the NPT, the rest of the world would have no illusions about Iranian behavior.

Contrast this series of events with a more likely path to an Iranian nuclear weapon: Iran operates a small, undeclared uranium enrichment facility to produce nuclear material beyond the gaze of IAEA inspectors. This would require transferring uranium feed material from Iran's existing enrichment plants or uranium conversion facility, or operating an undeclared uranium conversion facility. Diversions of uranium from declared facilities risk being noticed by inspectors, of course, but small quantities of material might be writ-

---

36. Fredrik Dahl, "Iran Submarine Plan May Fuel Western Nuclear Worries," *Reuters*, July 5, 2012, available from *www.reuters.com/article/2012/07/05/us-iran-nuclear-submarines-idUSBRE8640PC20120705*.

ten off under the broad category of "material unaccounted for," and in any case it probably would take the IAEA some time to pin down any discrepancies.[37] The end result may be the same—Iran would have nuclear material outside of safeguards with which to supply a nuclear weapons effort—but this pathway is likely to give Iran more time to actually develop nuclear weapons before the international community takes some action to delay or stop its program.

Iran is the illustrative case for the danger posed by the naval propulsion loophole, but even here it seems unlikely that Iran would actually take this route to a weapon. And if Tehran did choose to exempt nuclear material from safeguards and then divert it for weapons, this would probably be preferable—from the point of view of the international community—to a more covert strategy. Taking advantage of the naval propulsion loophole sounds a very clear alarm, giving the international community more time to bring both diplomatic and military pressure to bear to change Iran's course.

Mitigating the Naval Propulsion Loophole

There have been many proposals in recent years designed to strengthen the NPT and close its various loopholes, and the naval nuclear propulsion exemption has drawn its share of scrutiny. But before discussing policy options for heading off future nuclear submarine programs, we might ask whether closing this loophole is even worth the trouble. While the analysis above suggests that the naval propulsion exemption really does not pose much of a proliferation risk at the moment, the loophole becomes more dangerous once a precedent has been set that legitimizes the non-explosive

---

37. For a skeptical take on the IAEA's ability to quickly detect safeguards violations, see Henry D. Sokolski, "Assessing the IAEA's Ability To Verify the NPT," and Thomas B. Cochran, "Adequacy of IAEA's Safeguards for Achieving Timely Detection," in Henry D. Sokolski, ed., *Falling Behind: International Scrutiny of the Peaceful Atom*, Carlisle, PA: Strategic Studies Institute, 2008, pp. 3- 49, 121-157.

military use of nuclear material. The second state to use the exemption will have an easier time procuring sensitive technology, face less international pressure to change course, and generally set off fewer proliferation alarm bells. It makes sense, then, to try to hold the line and prevent countries from exercising the exemption in the first place or, if a precedent must be set, to try to limit its damage.

Several policy options are available, including voluntary safeguards agreements for naval nuclear reactors, efforts to limit the sale of nuclear submarine technology, a push to transition nuclear submarines to low enriched uranium fuel, and even a separate treaty on fissile material production. Most of these, however, do not really close the naval propulsion loophole; instead, they only reduce the likelihood that the loophole will be exercised, or render the loophole less dangerous if a state chooses to use it.

Safeguarding Naval Nuclear Propulsion

The decision by NPT drafters not to require safeguards on non-explosive military uses of nuclear material was very much a product of its time. States in the mid- to late-1960s were reluctant to allow inspectors access to civilian nuclear facilities, let alone military installations.[38] Safeguards concepts and techniques were not yet well established, and the mandate of the IAEA itself was undergoing a large shift as it prepared for the challenge of acting as the inspections body for the NPT. At the time of the NPT negotiations, IAEA inspections of military naval vessels or military-run reactors would have been seen as a substantial expansion of the agency's statutory authority.

The context for IAEA safeguards has changed significantly since those early days. IAEA inspectors have established on the Agency's behalf a reputation for fairness and discretion. The scope of the IAEA's safeguards work has expanded from merely verifying state

---

38. "Editorial Note," Document 92.

declarations and performing material accounting at select facilities; the Agency is now widely seen to be responsible for assuring that a member state has no undeclared nuclear activities. In the Iran case, the IAEA has even taken on the task of evaluating possible nuclear weapons-related activities that do not involve sensitive nuclear material.[39] Safeguards technology, too, has evolved, offering new tools for verifying non-diversion even when inspectors visit a site only infrequently.[40]

All of these changes mean that IAEA safeguards for naval reactors—a non-starter as recently as the late 1980s, when Canada was ready to take advantage of the naval propulsion exemption with only bilateral safeguards in place—now represent a reasonable approach to mitigating the proliferation risks of nuclear submarines. To be sure, nuclear submarines pose special challenges for existing safeguards techniques. A major selling point for nuclear propulsion, after all, is that nuclear submarines can venture further from port and stay away longer, and this complicates efforts to verify the non-diversion of material. Naval reactors may also pose technical

---

39. This last shift is not without some controversy. See, for example, Daniel Joyner, "Iran's Nuclear Program and the Legal Mandate of the IAEA," *JURIST*, November 9, 2011, available from *jurist.org/forum/2011/11/dan-joyner-iaea-report.php*. For a defense of the legal basis for the IAEA's activities in Iran, see David Albright, Olli Heinonen, and Orde Kittrie, *Understanding the IAEA's Mandate in Iran: Avoiding Misinterpretations*, Washington, DC: Institute for Science and International Security, November 27, 2012, available from *isis-online.org/uploads/isis-reports/documents/Misinterpreting_the_IAEA_27Nov2012.pdf*.

40. Important advances in safeguards technology include remote monitoring systems, commercial satellite imagery, and environmental sampling. See Mark Schanfein, "International Atomic Energy Unattended Monitoring Systems," in James E. Doyle, ed., *Nuclear Safeguards, Security, and Nonproliferation*, Boston: Elsevier, 2008, pp. 113-134; Kaluba Chitumbo, Stephen W. Robb, and John Hilliard, "Use of Commercial Satellite Imagery in Strengthening IAEA Safeguards," in Bhupendra Jasani and Gotthard Stein, eds., *Commercial Satellite Imagery: A Tactic in Nuclear Weapon Deterrence*, Chichester, UK: Praxis Publishing, 2002, pp. 23–48; and D.L. Donohue, "Strengthening IAEA Safeguards through Environmental Sampling and Analysis," *Journal of Alloys and Compounds*, Vols. 271–273, June 1998, pp. 11–18.

difficulties for verification because of their high burn-up rates and large quantities of fission products.[41] But these hurdles could be overcome sufficiently to at least provide belated notice of the diversion of nuclear material. Such a safeguards approach for naval propulsion has made it onto the IAEA's long-term research and development plan.[42] Fundamentally, naval nuclear propulsion safeguards of any kind would be a substantial improvement over the presumed solution today, which would exempt nuclear material from verification altogether. If implemented, voluntary monitoring of naval nuclear propulsion by the IAEA could go a long way toward mitigating the nonproliferation impact of a nuclear submarine program.

Convincing a state to subject its naval nuclear propulsion work to IAEA verification, however, may be a tough sell. Brazil, in particular, has illustrated its reluctance to take on new safeguards obligations by refusing to sign the Additional Protocol, even in the face of significant international pressure. Brazil may also be particularly sensitive to the risk that inspections would reveal information about military plans or capabilities. Brazil has cited concerns about the protection of proprietary technical information in seeking to limit IAEA inspector access to its centrifuge plant.[43] These concerns would probably be even more pronounced if inspections required that the IAEA gain access to a Brazilian military facility or the nuclear submarine itself.[44]

---

41. Mark E. Abhold, "Irradiated Fuel Measurements," in *Nuclear Safeguards, Security and Nonproliferation: Achieving Security with Technology and Policy*, p. 66.

42. International Atomic Energy Agency, *IAEA Department of Safeguards Long-Term R&D Plan*, 2012-2023, Vienna: International Atomic Energy Agency, January 2013, available from *https://www.bnl.gov/ISPO/docs/STR-375-IAEA-Safeguards-Long-Term-Plan.pdf*.

43. Squassoni and Fite.

44. On the other hand, Brazil's unique four-party safeguards arrangement (with Argentina, the IAEA, and the Brazilian-Argentine Agency for Accounting and Control of Nuclear Materials) could be interpreted as leaving an opening for

While some may resist voluntary monitoring, the international community does have at least one point of leverage in pressuring states to go along: The supply of nuclear submarine technology could be made contingent on this kind of alternative safeguards arrangement. Suppliers may also be in a position to dictate the use of low enriched uranium—less than five percent U-235—which would eliminate the justification for creating an infrastructure to enrich uranium at higher levels. France's deal with Brazil is not known to carry any kind of additional verification requirement, but a safeguards provision could be an important component of future agreements to supply nuclear submarine technology.

Limiting the Supply of Nuclear Submarine Technology

Another approach would seek to cut off the supply of nuclear submarine technology to new aspirant states. Given the small number of potential nuclear submarine supplier states—the P-5 weapons states and India—setting collective limits on the sale of naval nuclear propulsion technology, or stopping it completely, is at least a possibility.[45] Preventing the spread of nuclear submarines makes it harder for would-be proliferants to plausibly claim an exemption from safeguards and has the added bonus, as discussed above, of limiting demand for this capability; states are more likely to seek nuclear submarines when their rivals do the same. An agreement to limit supply might be negotiated within existing multilateral bodies—the Nuclear Suppliers Group is an obvious choice, although it currently excludes India—or it could be the focus of a separate

---

safeguards on nuclear material used in naval reactors. See Thielmann and Kelleher-Vergantini, pp. 6–7.

45. U.S. export controls already limit the participation of U.S. companies in foreign naval propulsion efforts under most circumstances. See Carlton E. Thorne, "Nonproliferation Export Controls," in *Nuclear Safeguards, Security and Nonproliferation: Achieving Security with Technology and Policy*, p. 541.

control regime.[46]

Limits on nuclear submarine technology transfer, however, probably would come too late to affect Brazil's nuclear submarine ambitions, as a supply agreement with France has already been concluded.[47] And restrictions on supply could not stop flagrant abuse of the naval propulsion loophole for nuclear weapons purposes; states could still claim a nuclear submarine program, produce highly enriched uranium, and divert it to a weapons effort. The absence of a known foreign supplier, however, might make the justification of a submarine program less believable, and thus make the exemption of nuclear material more clearly indicative of a nuclear weapons effort.

Transitioning to Low Enriched Uranium in Naval Reactors

The presence of a nuclear submarine program provides states with a built-in rationale for producing highly enriched uranium, which leads to greater proliferation risk. If a state opts to take full advantage of the loophole to support a weapons program, diversion of highly enriched uranium brings it that much closer to a nuclear weapon. Even if a state plays by the rules and uses the safeguards exemption only for naval propulsion, an infrastructure able to produce highly enriched uranium contributes to the state's latent nuclear capability and shortens the distance to a weapon should it ever decide to build one. Having highly enriched uranium around also complicates the state's nuclear security task and increases the risk of nuclear smuggling or sale to a third party. More generally, each additional state with highly enriched uranium has the effect of weakening global efforts to limit the production of sensitive nuclear materials. Control efforts, which largely rely on persuasion and bilateral cooperation agreements, can be undermined by the ability

---

46. Moltz, pp. 111–112.

47. Thielmann and Kelleher-Vergantini, pp. 4–5.

of states to point to others that have substantial enrichment and reprocessing capabilities.[48]

Most nuclear submarines burn highly enriched uranium fuel; uranium enriched to higher levels translates into a smaller reactor, longer operating periods, and less refueling. Modern naval nuclear reactors, however, can reasonably be powered by low enriched uranium. Only the United States and United Kingdom use weapons-grade nuclear material in their nuclear submarines, and China and France already use low enriched uranium to fuel their naval reactors.[49] If nuclear submarine aspirant countries could be convinced to adopt low enriched uranium as the fuel for their naval propulsion systems, this would help to mitigate some of the proliferation risk associated with these programs. There is room for some cautious optimism here. While an Iranian statement referring to "45-56 percent enriched uranium" for naval propulsion has drawn some attention, Brazil plans to use low enriched uranium for its nuclear submarines.[50] States that already field nuclear submarines running on highly enriched fuel—particularly the United States and United Kingdom—could help matters by considering a transition to low enriched uranium fuel for their naval propulsion programs. A recent U.S. Department of Energy report found such a transition would be feasible but uneconomical; this at least leaves the door open to a policy determination that the added expense might be worth it to realize broader nonproliferation goals.[51]

---

48. On the importance of international precedent in efforts to limit the spread of enrichment and reprocessing capabilities, see Jeffrey M. Kaplow and Rebecca Davis Gibbons, *The Days After a Deal with Iran: Implications for the Nuclear Nonproliferation Regime*, RAND Perspectives, Washington, DC: RAND Corporation, 2015, available from *www.rand.org/content/dam/rand/pubs/perspectives/ PE100/PE135/RAND_PE135.pdf*.

49. Ma and von Hippel, p. 91 and Thielmann and Kelleher-Vergantini, p. 2.

50. Thielmann and Kelleher-Vergantini, p. 2 and "Iran Mulls Highly Enriched Uranium."

51. Office of Naval Reactors, *Report on Low Enriched Uranium for Naval Re-*

Closing the Loophole with Legal Obligations

The NPT does have an amendment procedure, laid out in Article VIII of the treaty, but it is not of much practical use. Amendments do not take effect without the ratification of the five nuclear weapons states recognized by the NPT, all the members of the IAEA Board of Governors, and a majority of member states. An addition to the treaty still would not be binding for member states until they themselves ratify the amendment, and so for existing members an amendment would not amount to much more than a voluntary obligation that they could choose to take on. It would make more sense, then, to try to fill the naval propulsion loophole as part of a broader control treaty. There is some precedent for using new treaties to plug loopholes in the NPT: the Comprehensive Nuclear-Test-Ban Treaty largely closes the NPT loophole allowing non-nuclear weapons states to benefit from research into "peaceful nuclear explosions." A peaceful nuclear explosion—for example, using a nuclear blast to excavate a canal—is technically identical to a nuclear weapons test.

The best candidate for an international agreement to fill the naval propulsion loophole is the Fissile Material Cut-off Treaty (FMCT). Existing proposals for the FMCT would stop the production of highly enriched uranium or plutonium for nuclear weapons, but the treaty could be extended in negotiations to cover the production of nuclear material for naval propulsion as well.[52] The FMCT is not on a fast track, however. First taken up for negotiations at the United Nations (UN) Conference on Disarmament in 1995, the treaty has languished in a body that operates by consensus. In recent years, Pakistan has been the primary impediment, blocking negotiations even as it adds to its own stocks of fissile material.[53] Still, there are

---

actor Cores, Report to Congress, Washington, DC: U.S. Department of Energy, January 2014.

52. Ma and von Hippel, p. 87.

53. Peter Crail, "Pakistan's Nuclear Buildup Vexes FMCT Talks," *Arms Con-*

some signs that talks on the FMCT may finally be moving forward, as discussions have shifted to a smaller group within the Conference on Disarmament that excludes Pakistan.[54] Progress on this treaty may bring the international community closer to a real solution for the naval propulsion loophole.

Conclusion

Among the NPT's various shortcomings, the naval propulsion loophole stands out. It was created by a gap in treaty coverage, rather than by explicit language. The NPT's drafters intentionally omitted language on military non-explosive uses of nuclear technology, with the full understanding that it fashioned a loophole that might be exploited by states seeking nuclear weapons. The nuclear submarine exemption probably encouraged key states to join the treaty, however, and requires states to make a declaration before removing nuclear material from safeguards. That the naval propulsion exemption today is highly alerting—if exercised by some states, it probably would be a fairly good indicator of a nuclear weapons program—partly mitigates the proliferation risk of the loophole. Use of the exemption would be more alarming, however, because no state has ever taken advantage of it. Once a precedent is set for exempting material from safeguards, the loophole becomes less costly for states to employ and thus a greater proliferation risk. For this reason, the international community has an incentive to mitigate the proliferation consequences of nuclear submarine programs.

While Iran has drawn attention recently for hinting at nuclear submarine ambitions, Brazil is the real contender for the state most

---

trol Today, Vol. 41, No. 2, March 2011, available from *www.armscontrol.org/act/2011_03/Pakistan.*

54. Michael Krepon, "Will Pakistan and India Break the Fissile Material Deadlock?" *Arms Control Wonk* (blog), July 31, 2014, available from *krepon.armscontrolwonk.com/archive/4217/fmct.*

likely to set a dangerous nuclear submarine precedent. The Brazil case is difficult and complex; Brazil has long refused to bow to international pressure on nonproliferation matters, and its nuclear submarine effort has its origins in the military's nuclear weapons program. But Brazil also has no interest in throwing open the door to unsafeguarded nuclear material in states like Iran. The international community—and particularly France, which is supplying Brazil with submarine technology—has an opening to convince Brazil that it should not take advantage of the naval nuclear propulsion exemption. Rather, it should negotiate a supplementary safeguards approach with the IAEA that maintains some level of assurance against non-diversion of nuclear material. A voluntary safeguards agreement, if implemented, would significantly reduce the proliferation impact of Brazil's nuclear submarine effort.

This chapter's analysis of the naval nuclear propulsion loophole speaks to a broader issue with the way we evaluate the effectiveness of international legal constraints. The NPT is in some ways a victim of its own success. As the undisputed lynchpin of the nuclear nonproliferation regime, the NPT is a magnet for criticism, and, indeed, the treaty has a number of significant loopholes. It does not follow, however, as some analysts have suggested, that the NPT's gaps leave it wholly ineffective or even harmful in its own right to nonproliferation goals. Some of the treaty's flaws, to the extent that they encourage additional state adherence and provide information about potential noncompliance, may even be a source of strength.

Chapter 5

Another Gap in the NPT: How Israel and Others Get Outside Nuclear Help

Victor Gilinsky

Germany's supply to Israel of advanced submarines designed to launch long-range nuclear cruise missiles exemplifies a gap in the international effort to control the spread of nuclear weapons. There are other examples of this problem, involving other sets of countries, but this is the clearest one.

Germany is a member of the Nonproliferation Treaty (NPT), but the Treaty, routinely described as the "cornerstone" of the so-called nonproliferation regime, does not cover this kind of transaction.[1] Nor is there any other mechanism analogous to, say, the Nuclear Suppliers Group that deals with worrisome civil nuclear energy transactions, to control it.

The Treaty prohibits the five "legitimate" nuclear-weapon states—the United States, Russia, Britain, China, and France—from giving nuclear weapons to *any* other states, or to "assist, encourage,

---

1. The first sentence on the State Department NPT web page is: "The Treaty on the Nonproliferation of Nuclear Weapons is the cornerstone of the nonproliferation regime." U.S. Department of State, "Nuclear Nonproliferation Treaty," available from *http://www.state.gov/t/isn/npt/*, accessed on February 23, 2016. The UN's web page on the 2015 NPT Review Conference states: "Since its entry into force, the NPT has been the cornerstone of global nuclear non-proliferation regime." United Nations, "2015 NPT Review Conference," available from *http://www.un.org/en/conf/npt/2015/*, accessed on February 23, 2016.

or induce" any such state to get such weapons.[2] The other Treaty members may not receive or make weapons, or seek or obtain assistance to make them, but the Treaty does not prohibit them from helping countries that are not Treaty members with their nuclear weapons programs.

There was a certain logic to this formulation. The Treaty drafters were narrowly fixed on controlling actual warheads, as opposed to, say, delivery vehicles. If a Treaty member aside from the five weapon states had nuclear weapon technology to offer others, it would already have been in violation of the Treaty, and so an additional provision covering this possibility was unnecessary. The trouble is, this narrowly-focused approach on what is impermissible reflects an overly simplistic view of the danger of nuclear weapon spread and what it takes to prevent it.

On the civil nuclear power side, the main technology suppliers recognized after the 1974 Indian bomb explosion that international security required a degree of control over nuclear energy technology transfers—in particular those relating to production of fuels that are also nuclear explosives—to supplement the narrowly drafted prohibitions of the NPT. More recently, on the weapon side, a 2004 United Nations Security Council Resolution acknowledged that the dangers of nuclear proliferation were exacerbated by more than transfer of warheads and directly related technology. It added the "spread of the means of delivery" to the items that "constitute a threat to international peace and security."[3] The Security Council

---

2. Article 1 of the Nonproliferation Treaty, International Atomic Energy Agency, "Treaty on the Non-Proliferation of Nuclear Weapons," April 22, 1970, p. 2.

3. Resolution 1540 (2004) Adopted by the Security Council on April 28, 2004. The NPT preamble recognizes that nuclear weapons and means of delivery go together by coupling the need to eliminate "means of delivery" of nuclear weapons with eliminating the weapons themselves. It seems reasonable to read this as an intermediate step toward the distant goal of general and complete disarmament, rather than something that had to wait for the lion to lie with the lamb:
> Desiring to further the easing of international tension and the strengthening of trust between States in order to facilitate the

made the point again in a 2009 Resolution, which while primarily dealing with threats from non-state groups, reaffirmed the general proposition that "proliferation of weapons of mass destruction, and their means of delivery, constitutes a threat to international peace and security."[4]

The Security Council's definition of "means of delivery" lists missiles and rockets and the like, rather than planes or ships.[5] But the German-supplied subs are specifically designed with extra-large torpedo tubes to be the firing mechanism for the nuclear-tipped cruise missiles. Submarine and missile are integrally connected in terms of hardware and mission, so it is reasonable to include the German-supplied so-called Dolphin-class submarines in the "means of delivery" category.

The procurement process started in the late 1980s. Israel first contemplated construction in a U.S. naval yard but turned to Germany when that country agreed to pay for the first two submarines. Germany's position vis-à-vis Israel had become especially awkward after the first Gulf War when it came to light that German firms had helped Saddam's missile program. Some 35 such missiles reached Israel. Germany's contribution was cast as a continuation of the reparations process for the WWII murder of millions of Jews.[6]

---

cessation of the manufacture of nuclear weapons, the liquidation of all their existing stockpiles, and the elimination from national arsenals of nuclear weapons and the means of their delivery pursuant to a Treaty on general and complete disarmament under strict and effective international control.

See United Nations Security Council, "Resolution 1540 (2004)," April 28, 2004.

4. United Nations Security Council, "Resolution 1887 (2009)," September 24, 2009.

5. United Nations Security Council, "Resolution 1540 (2004)," April 28, 2004.

6. Alona Ferber and Judy Maltz, "The Surprising Story Behind Israel's Complicated Love Affair With Germany," *Haaretz*, May 12, 2015, available from *http://www.haaretz.com/israel-news/.premium-1.655332*.

Chapter 5   157

The connection of the submarines with a nuclear mission should have been clear to Germany from the start of the procurement process in 1990, but in any case not much later. Israel was clearly motivated by its perceived need to respond to what seemed to it an impending Iranian nuclear weapon. Other countries, including the United States, could also not have missed what was going on. The information about the submarines' nuclear mission has been reported in the Israeli press (with suitable qualifications that the information comes from foreign sources) at least since the late 1990s, and increasingly so.[7] To jump to recent years, a 2011 story on *Ynetnews.com* reports an interview with the submarine fleet's commander, who is said to be "privy to the State of Israel's deepest secrets."[8] The article headline was, "Doomsday weapon: Israel's

---

7. Ha'aretz June 9, 1998, stated: At the beginning of 1999, when the navy will bring into active service the first of three Dolphin submarines constructed at German shipyards, the Middle East arms race will take on new proportions. [The reference is to the expectation at the time, based on Israeli intelligence estimates, that Iran would soon obtain nuclear weapons.] Yossi Melman, "Swimming with the Dolphins," *Ha'aretz*, June 9, 1998, available from *http://fas.org/nuke/guide/israel/sub/internatl1-1.html*. A July 1, 1998, Washington Times story, "Israel buying 3 submarines to carry nuclear missiles," stated, "Israel is buying three large submarines from Germany capable of carrying nuclear-armed cruise missiles, with the reported goal of deterring any enemy [Iran] from trying to take out its nuclear weapons with a surprise attack." Similar accounts appeared in the Arab press. Martin Sieff, "Israel buying 3 submarines to carry nuclear missiles," *The Washington Times*, July 1, 1998, available from *http://fas.org/nuke/guide/israel/sub/internatl1.html*.; Recently, from Haaretz, September 23, 2014, "Israel's Fourth Dolphin-Class Submarine Docks at Haifa":

> According to reports in foreign media, the German-made submarines can carry cruise missiles with a range of thousands of kilometers, and can be equipped with nuclear warheads. According to these reports, the Israeli submarine fleet is meant to allow for a "second strike" in the event of a nuclear attack.

"Israel's Fourth Dolphin-class Submarine Docks in Haifa," *Haaretz*, September 23, 2014, available from *http://www.haaretz.com/israel-news/.premium-1.617543*.

8. Alex Fishman, "Doomsday weapon: Israel's submarines," *Ynetnews*, October 9, 2011, available from *http://www.ynetnews.com/articles/0,7340,L-4120185,00.html*.

submarines." It left little to the imagination.

The subject has been aired in the German press. A 2012 series on the subject in *Der Spiegel* put it this way:

> Research SPIEGEL has conducted in Germany, Israel and the United States, among current and past government ministers, military officials, defense engineers and intelligence agents, no longer leaves any room for doubt: With the help of German maritime technology, Israel has managed to create for itself a floating nuclear weapon arsenal: submarines equipped with nuclear capability.[9]

It included the following:

> Insiders say that the Israeli defense technology company Rafael built the missiles for the nuclear weapons option. Apparently it involves a further development of cruise missiles of the Popeye Turbo SLCM type, which are supposed to have a range of around 1,500 kilometers (940 miles) and which could reach Iran with a warhead weighing up to 200 kilograms (440 pounds).

Despite these reports, the German government has stuck to its position that it knew nothing about an Israeli nuclear weapons program (as does the U.S. government). German Chancellor Angela Merkel has repeatedly said she feels a special obligation to Israel's security, in light of the Holocaust committed by the Nazis.[10]

---

9. "Operation Samson: Israel's Deployment of Nuclear Missiles on Subs from Germany," *Der Spiegel*, June 4, 2012, available from *http://www.spiegel.de/international/world/israel-deploys-nuclear-weapons-on-german-built-submarines-a-836784.html*.

10. For example, in 2008 when Chancellor Merkel addressed the Israeli parliament she stated "Israel's security is never negotiable." See Alona Ferber and Judy Maltz.

Any shred of doubt about Israel's possession of nuclear weapons, and in particular about the presence of long-range nuclear missiles on the German-supplied submarines, got erased at the January 12 ceremony celebrating the arrival in Haifa of the *Rahav*, the fifth of six submarines to come from its German shipyard. The *Rahav* is a highly advanced diesel-electric boat that in certain respects is superior to nuclear-propelled ones. The three most modern Dolphins are equipped with air-independent propulsion—they carry their own oxygen supply—and so can stay beneath the surface for weeks. They are quieter than nuclear submarines.

Israel relaxed it otherwise extremely tight censorship over nuclear weapon deployment precisely because its long-range nuclear weapons are no longer weapons of last resort, to be used only *in extremis*; they are now Israel's deterrent force, integrated into its overall strategy. At the January 12 ceremony, Israeli Prime Minister Benjamin Netanyahu said the "submarine fleet is used first and foremost to deter our enemies who strive to extinguish us. . . .They must know that Israel is capable of hitting back hard against anyone who seeks to hurt us. . . Israel's citizens need to know that it is a very strong state."[11]

For a deterrent to work, the antagonist (read, Iran) has to be aware of it, ergo the nuclear force has to be publicized, even flaunted. That the word "nuclear" is left out doesn't detract from the point, one that no one can miss. The omission highlights an advantage to Israel of its so-called opacity policy. Such is the nature of human psychology that advertising its nuclear weapons, while omitting the word "nuclear," both puts adversaries on notice and allows Israel's suppliers and supporters to maintain their hypocritical stance. If Prime Minister Netanyahu blurted out the truth, very likely Germany could not continue to supply Israel with submarines intended to carry nuclear weapons.

---

11. Ofira Koopmans, "Israel's Fifth German Submarine Arrives at Haifa Port," *Haaretz*, January 12, 2016, available from *http://www.haaretz.com/israel-news/1.697061*.

Israel doesn't similarly publicize its ground-based nuclear missiles because it fears they may be vulnerable to ground or missile attack, whereas the submarines are securely hidden in the ocean. There isn't much doubt at whom the sea-based missiles would be pointed at: For years Prime Minister Netanyahu has been warning that Iran is intent on getting nuclear weapons that it intends to use against Israel.[12]

The long and short of this account is that the German government, which paid for a good part of the cost of the submarines, has not only known their real mission, but supported it deliberately. As *Der Spiegel* put it:

> The German government has always pursued an unwritten rule on its Israel policy, which has already lasted half a century and survived all changes of administrations, and that former Chancellor Gerhard Schröder summarized in 2002 when he said: "I want to be very clear: Israel receives what it needs to maintain its security.[13]

What are we to make of this? Supplying the submarines, even knowing their primary mission was to be platforms for nuclear weapons, is not itself a violation of the NPT because, as we have seen, the Treaty does not put restrictions on supply of weapon-related technology or materials from non-nuclear-weapon states. Nevertheless, supplying an NPT holdout, even one with historical claims, with the critical delivery vehicles for its nuclear force would seem to violate the spirit of the Treaty.[14]

---

12. For example, Netanyahu's January 2015 remark: "The ayatollahs in Iran, they deny the Holocaust while planning another genocide against our people."
Peter Beinart, "Iran Is Not an 'Existential' Threat to Israel - No Matter What Netanyahu Claims," *Haaretz*, August 7, 2015, available from *http://www.haaretz.com/opinion/1.670097*.

13. "Operation Samson," *Der Spiegel*.

14. Germany also had to have known that Israel illegally slipped 200 tons of

There is something ludicrous about a nonproliferation regime that prevents Germany, among other technology suppliers, from providing Israel with enrichment or reprocessing technology, or even a power reactor, but permits it to supply integral components of Israel's strategic nuclear forces.

The U.S. government (USG) has obviously been aware of the German-Israeli sub deal and what it was really about and has been silent on it. Like the German government, the USG pretends it knows nothing about any nuclear weapons in Israel. At the same time, it has done everything it can diplomatically to protect Israel from any criticism, or in fact, any inquiry, on this subject. It may have done more. The U.S. Army Corps of Engineers' Europe District has done extensive military and naval construction in Israel, including at the Haifa Navy Base, the homeport of the Dolphin-class submarines. The Europe District maintains a Project Office at the Haifa Navy Base.[15] It would not be surprising if the Army Engineers did work on the Dolphin-class submarines' docks.

While Germany's supply of submarines for nuclear missions may strictly speaking be permissible under the NPT rules, any U.S. participation in Israel's nuclear weapon activities falls in a different category. The United States is subject to the strictures of the Treaty's Article I, under which it undertakes "not in any way to assist, encourage, or induce any non-nuclear-weapon State to manufacture or otherwise acquire nuclear weapons . . . ."[16] The truth is that

---

uranium out of the European Community in 1968 in what has become known as Operation Plumbat.

15. It is one of three U.S. Army Corps of Engineers Europe District Project Offices in Israel.

16. Article I applies to the five nuclear-weapon states, and yet Russia assists India to develop a nuclear submarine force that includes ballistic missile subs, China assists Pakistan's nuclear weapon program, and the United States assisted India's nuclear weapon program by arranging for it the largest gift of all—a waiver for India from the nuclear trade sanctions imposed because of its refusal to join the NPT, and in fact its decades long opposition to it. *Quis custodiet ipsos custodes?*

while U.S. proclaimed policy is that *all* countries should become members of the NPT (which they would have to do as non-nuclear-weapon states), the real policy is different.

Any issue relating to Israel is heavily laden with U.S. domestic political considerations. Touching on Israel's nuclear forces is in Washington the no-no of no-noes. Everyone in Washington understands that it is no way to advance one's career. The U.S. government does not even acknowledge the existence of Israeli nuclear weapons and refuses to discuss the subject, apparently even internally on a classified basis. The United States has consistently protected Israel's nuclear monopoly in the Middle East, in part by vetoing efforts by other countries to raise the subject in international arenas. An argument in favor of this approach is that given that Israel has nuclear weapons, it is better that they be secure, to avoid situations in which Israel might be tempted to use them for fear of losing them. But there is another side to the argument.

Israel, and in fact all the NPT holdouts—India, Israel, North Korea, and Pakistan—are the most likely countries to use nuclear weapons against their adversaries.[17] All four are involved in bitter disputes. While they all speak of using their weapons for deterrence, they do not rule out use of the weapons in response to non-nuclear provocation. For example, Pakistan is now boasting of having introduced a class of battlefield weapons, which they intend to use to ward off Indian incursions into Pakistani territory (which India threatened in response to its claim of Pakistani-inspired terrorism on Indian territory).[18] Israel describes its nuclear force (omitting

---

17. In 1969, in considering the possibility of Israeli nuclear weapons, Henry Kissinger wrote President Nixon about the danger—that if the Israelis had them (which they were in the process of doing), they would be the most likely country to use them. David Stout, "Israel's Nuclear Arsenal Vexed Nixon," *The New York Times*, November 29, 2007, available from *http://www.nytimes.com/2007/11/29/world/middleeast/29nixon.html?_r=0*.

18. Ankit Panda, "Pakistan Clarifies Conditions for Tactical Nuclear Weapon Use Against India," *The Diplomat*, October 20, 2015, available from *http://thediplomat.com/2015/10/pakistan-clarifies-conditions-for-tactical-nuclear-weapon-use-*

"nuclear," of course), at least the sea-based leg of its triad, as a secure second-strike force. But a second-strike force in a tiny country that can be effectively eliminated by one nuclear weapon is a very different thing than such a force in a country with strategic depth. One has the impression that Israel's second-strike force is a very forward-leaning one, and that in Israeli thinking its "second strike" will arrive before the adversary's first one, and possibly before its adversary even has the wherewithal for a first one. It makes for a dangerous state of affairs.

There are depths below depths in the nuclear weapons world and countries that have some nuclear weapons may in time get a great many. Those with tens may get hundreds, and those with hundreds could decide to get thousands. That may not make a lot of sense, but our Cold War experience should guard us against optimism on this score. The work of non-proliferation regarding the NPT holdouts should not stop because they have nuclear weapons. We should not give up on constraining these nuclear weapon programs.

Insofar as Israel is concerned, the most effective step in this direction, and one without which no progress is possible, is to force the U.S. government and European Community to acknowledge Israel's nuclear weapons. Forcing democratic governments to end their pretense would lance the current policy of pretending to support universal application of the NPT but at the same time engaging in trade and practices that undermine the Treaty. The entire world is aware of this hypocrisy, resulting in a cynical view of the so-called nonproliferation regime.

There remains the more general problem created by the gap in application of the NPT—the lack of a prohibition on non-nuclear-weapon Treaty members from supplying other states with essential components of nuclear weapon systems. These other states could be non-members, as in the example covered in this paper, but they could also be member states with apparent nuclear weapon am-

---

against-india/.

bitions. Amending Article II of the Treaty is of course out of the question. But one could contemplate an organization ancillary to the Treaty, perhaps one analogous to the Nuclear Suppliers Group, before which specific weapon technology transfers—primarily technology related to nuclear weapons delivery—would be brought for discussion and resolution. Above all, we should not give in to the world-weary sophistication that there is nothing to be done.

Chapter 6

Locking Down the NPT[1]

Henry D. Sokolski and Victor Gilinsky

As President Donald Trump considers how his administration will prevent the further proliferation of nuclear weapons, it is useful to note that every president since Lyndon B. Johnson has spoken of the importance of the Nuclear Nonproliferation Treaty (NPT). Yet, at the same time, they have all, to a lesser or greater degree, weakened the treaty, through lax enforcement, by carving out exceptions for certain countries, or by just ignoring it. We have come to the point now that North Korea, which signed the treaty in 1985, is now mocking it. And in all the discussions over a possible Iranian bomb, no one seems to think the treaty's 90-day withdrawal clause would be much of a hurdle if Tehran decided to leave the NPT.

If President Trump really wants to strengthen the treaty, a good—and necessary—place to start is to make it much more difficult for any of the 189 member states to leave the NPT. It is at odds with the NPT's purpose to allow a country to import or develop technology under the treaty's cover and then walk out to make bombs. At a minimum, before legally exiting the treaty, a country should have to clear its NPT obligations by returning whatever it got from others based on the understanding that it was a good-faith treaty member.

The background of the North Korean bomb is instructive. The In-

---

1. A version of this chapter was originally published in the Bulletin of Atomic Scientists, June 17, 2009, available from *thebulletin.org/locking-down-npt*.

ternational Atomic Energy Agency (IAEA), of which the United States is the most prominent member, allowed North Korea to drag out its obligation to undergo a thorough initial IAEA inspection within 18 months of signing and ratifying the NPT. The inspection did not start until 1992, by which time Pyongyang already had illicitly separated plutonium. When the inspectors insisted on inspecting two waste sites that might reveal this, the North Koreans tossed them out and threatened to pull out of the treaty altogether.

The international reaction to this behavior was not to treat North Korea as an NPT violator but to beseech it to remain in the treaty. The United States went so far as to offer two large light water reactors (with South Korea and Japan footing the $5 billion bill), and to agree to shield the North for years from the NPT's inspection requirements in return for a halt in North Korean plutonium production. When it later looked as if Washington would ultimately insist on inspection, the North announced it was withdrawing from the treaty anyhow. This led to lots of diplomatic hand wringing in NPT-signatory state capitals, but not a peep that Pyongyang couldn't legally quit the NPT.

Such a permissive interpretation of the withdrawal clause must change to one that conforms to the treaty's purpose--no matter how awkward or late in the game. The international community should insist that North Korea is still an NPT member and that, among its other obligations, it must permit the basic IAEA inspections that never took place in 1992.

The predictable reaction to this proposal, in some quarters, will be that North Korea's actions can't be reversed by international pressure, and that the NPT's withdrawal clause, which was made deliberately permissive to entice prospective member states, can't be reinterpreted. Yet, the United States has long been committed to reversing North Korea's nuclear status, and if we don't reinterpret the treaty we will be writing it off as a serious commitment. International legal obligations do matter and being branded as a treaty violator has consequences internationally. Even Pyongyang seems to sense this and keeps repeating that it acted legally in leaving the NPT. For example, in its June 14, 2009, statement defending its right to test nuclear weapons it insisted, "[North Korea's] second nuclear test . . . does not run counter to any international law."

To make a new approach stick we will need the broad support of NPT members. But that won't happen until the United States steps forward to announce a new and tougher standard for withdrawal. This might not be welcomed by those self-styled realists who believe insisting on strict NPT compliance gets in the way of reaching accommodations with difficult countries. But a permissive approach didn't work with North Korea and has done great harm to the treaty overall.

Which brings us to Iran: Tehran imported considerable nuclear technology as a treaty member, most prominently from Russia for its Bushehr nuclear power plant project. We should make clear now that if it, or any other country, chooses to withdraw from the NPT, they would be obligated to return or cease using such imports before they could clear their treaty accounts. Lacking that, they would be international outlaws. Raising the bar to withdrawal is an essential first step in strengthening the treaty to deter would-be bomb makers.

Appendix

The Treaty on the Non-Proliferation of Nuclear Weapons (NPT)

The States concluding this Treaty, hereinafter referred to as the Parties to the Treaty,

Considering the devastation that would be visited upon all mankind by a nuclear war and the consequent need to make every effort to avert the danger of such a war and to take measures to safeguard the security of peoples,

Believing that the proliferation of nuclear weapons would seriously enhance the danger of nuclear war,

In conformity with resolutions of the United Nations General Assembly calling for the conclusion of an agreement on the prevention of wider dissemination of nuclear weapons,

Undertaking to co-operate in facilitating the application of International Atomic Energy Agency safeguards on peaceful nuclear activities,

Expressing their support for research, development and other efforts to further the application, within the framework of the International Atomic Energy Agency safeguards system, of the principle of safeguarding effectively the flow of source and special fissionable materials by use of instruments and other techniques at certain strategic points,

Affirming the principle that the benefits of peaceful applications of nuclear technology, including any technological by-products which may be derived by nuclear-weapon States from the development of nuclear explosive devices, should be available for peaceful purposes to all Parties to the Treaty, whether nuclear-weapon or non-nuclear-weapon States,

Convinced that, in furtherance of this principle, all Parties to the Treaty are entitled to participate in the fullest possible exchange of scientific information for, and to contribute alone or in co-operation with other States to, the further development of the applications of atomic energy for peaceful purposes,

Declaring their intention to achieve at the earliest possible date the cessation of the nuclear arms race and to undertake effective measures in the direction of nuclear disarmament,

Urging the co-operation of all States in the attainment of this objective,

Recalling the determination expressed by the Parties to the 1963 Treaty banning nuclear weapons tests in the atmosphere, in outer space and under water in its Preamble to seek to achieve the discontinuance of all test explosions of nuclear weapons for all time and to continue negotiations to this end,

Desiring to further the easing of international tension and the strengthening of trust between States in order to facilitate the cessation of the manufacture of nuclear weapons, the liquidation of all their existing stockpiles, and the elimination from national arsenals of nuclear weapons and the means of their delivery pursuant to a Treaty on general and complete disarmament under strict and effective international control,

Recalling that, in accordance with the Charter of the United Nations, States must refrain in their international relations from the threat or use of force against the territorial integrity or political independence of any State, or in any other manner inconsistent with the Purposes of the United Nations, and that the establishment and maintenance of international peace and security are to be promoted with the least diversion for armaments of the world's human and economic resources,

Have agreed as follows:

Article I

Each nuclear-weapon State Party to the Treaty undertakes not to transfer to any recipient whatsoever nuclear weapons or other nuclear explosive devices or control over such weapons or explosive devices directly, or indirectly; and not in any way to assist, encourage, or induce any non-nuclear-weapon State to manufacture or otherwise acquire nuclear weapons or other nuclear explosive devices, or control over such weapons or explosive devices.

Article II

Each non-nuclear-weapon State Party to the Treaty undertakes not to receive the transfer from any transfer or whatsoever of nuclear weapons or other nuclear explosive devices or of control over such weapons or explosive devices directly, or indirectly; not to manufacture or otherwise acquire nuclear weapons or other nuclear explosive devices; and not to seek or receive any assistance in the manufacture of nuclear weapons or other nuclear explosive devices.

Article III

1. Each non-nuclear-weapon State Party to the Treaty undertakes to accept safeguards, as set forth in an agreement to be negotiated and concluded with the International Atomic Energy Agency in accordance with the Statute of the International Atomic Energy Agency and the Agency's safeguards system, for the exclusive purpose of verification of the fulfilment of its obligations assumed under this

Treaty with a view to preventing diversion of nuclear energy from peaceful uses to nuclear weapons or other nuclear explosive devices. Procedures for the safeguards required by this Article shall be followed with respect to source or special fissionable material whether it is being produced, processed or used in any principal nuclear facility or is outside any such facility. The safeguards required by this Article shall be applied on all source or special fissionable material in all peaceful nuclear activities within the territory of such State, under its jurisdiction, or carried out under its control anywhere.

2. Each State Party to the Treaty undertakes not to provide: (a) source or special fissionable material, or (b) equipment or material especially designed or prepared for the processing, use or production of special fissionable material, to any non-nuclear-weapon State for peaceful purposes, unless the source or special fissionable material shall be subject to the safeguards required by this Article.

3. The safeguards required by this Article shall be implemented in a manner designed to comply with Article IV of this Treaty, and to avoid hampering the economic or technological development of the Parties or international co-operation in the field of peaceful nuclear activities, including the international exchange of nuclear material and equipment for the processing, use or production of nuclear material for peaceful purposes in accordance with the provisions of this Article and the principle of safeguarding set forth in the Preamble of the Treaty.

4. Non-nuclear-weapon States Party to the Treaty shall conclude agreements with the International Atomic Energy Agency to meet the requirements of this Article either individually or together with other States in accordance with the Statute of the International Atomic Energy Agency. Negotiation of such agreements shall commence within 180 days from the original entry into force of this Treaty. For States depositing their instruments of ratification or accession after the 180-day period, negotiation of such agreements shall commence not later than the date of such deposit. Such agree-

ments shall enter into force not later than eighteen months after the date of initiation of negotiations.

Article IV

1. Nothing in this Treaty shall be interpreted as affecting the inalienable right of all the Parties to the Treaty to develop research, production and use of nuclear energy for peaceful purposes without discrimination and in conformity with Articles I and II of this Treaty.

2. All the Parties to the Treaty undertake to facilitate, and have the right to participate in, the fullest possible exchange of equipment, materials and scientific and technological information for the peaceful uses of nuclear energy. Parties to the Treaty in a position to do so shall also co-operate in contributing alone or together with other States or international organizations to the further development of the applications of nuclear energy for peaceful purposes, especially in the territories of non-nuclear-weapon States Party to the Treaty, with due consideration for the needs of the developing areas of the world.

Article V

Each Party to the Treaty undertakes to take appropriate measures to ensure that, in accordance with this Treaty, under appropriate international observation and through appropriate international procedures, potential benefits from any peaceful applications of nuclear explosions will be made available to non-nuclear-weapon States Party to the Treaty on a non-discriminatory basis and that the charge to such Parties for the explosive devices used will be as low as possible and exclude any charge for research and development. Non-nuclear-weapon States Party to the Treaty shall be able to obtain such benefits, pursuant to a special international agreement or agreements, through an appropriate international body with adequate representation of non-nuclear-weapon States. Negotiations on this subject shall commence as soon as possible after the Treaty enters into force. Non-nuclear-weapon States Party to the Treaty so desiring may also obtain such benefits pursuant to bilateral agreements.

Article VI

Each of the Parties to the Treaty undertakes to pursue negotiations in good faith on effective measures relating to cessation of the nuclear arms race at an early date and to nuclear disarmament, and on a treaty on general and complete disarmament under strict and effective international control

Article VII

Nothing in this Treaty affects the right of any group of States to conclude regional treaties in order to assure the total absence of nuclear weapons in their respective territories.

Article VIII

1. Any Party to the Treaty may propose amendments to this Treaty. The text of any proposed amendment shall be submitted to the Depositary Governments which shall circulate it to all Parties to the Treaty. Thereupon, if requested to do so by one-third or more of the Parties to the Treaty, the Depositary Governments shall convene a conference, to which they shall invite all the Parties to the Treaty, to consider such an amendment.

2. Any amendment to this Treaty must be approved by a majority of the votes of all the Parties to the Treaty, including the votes of all nuclear-weapon States Party to the Treaty and all other Parties

which, on the date the amendment is circulated, are members of the Board of Governors of the International Atomic Energy Agency. The amendment shall enter into force for each Party that deposits its instrument of ratification of the amendment upon the deposit of such instruments of ratification by a majority of all the Parties, including the instruments of ratification of all nuclear-weapon States Party to the Treaty and all other Parties which, on the date the amendment is circulated, are members of the Board of Governors of the International Atomic Energy Agency. Thereafter, it shall enter into force for any other Party upon the deposit of its instrument of ratification of the amendment.

3. Five years after the entry into force of this Treaty, a conference of Parties to the Treaty shall be held in Geneva, Switzerland, in order to review the operation of this Treaty with a view to assuring that the purposes of the Preamble and the provisions of the Treaty are being realised. At intervals of five years thereafter, a majority of the Parties to the Treaty may obtain, by submitting a proposal to this effect to the Depositary Governments, the convening of further conferences with the same objective of reviewing the operation of the Treaty.

Article IX

1. This Treaty shall be open to all States for signature. Any State which does not sign the Treaty before its entry into force in accordance with paragraph 3 of this Article may accede to it at any time.

2. This Treaty shall be subject to ratification by signatory States. Instruments of ratification and instruments of accession shall be deposited with the Governments of the United Kingdom of Great Britain and Northern Ireland, the Union of Soviet Socialist Republics and the United States of America, which are hereby designated the Depositary Governments.

3. This Treaty shall enter into force after its ratification by the

States, the Governments of which are designated Depositaries of the Treaty, and forty other States signatory to this Treaty and the deposit of their instruments of ratification. For the purposes of this Treaty, a nuclear-weapon State is one which has manufactured and exploded a nuclear weapon or other nuclear explosive device prior to 1 January 1967.

4. For States whose instruments of ratification or accession are deposited subsequent to the entry into force of this Treaty, it shall enter into force on the date of the deposit of their instruments of ratification or accession.

5. The Depositary Governments shall promptly inform all signatory and acceding States of the date of each signature, the date of deposit of each instrument of ratification or of accession, the date of the entry into force of this Treaty, and the date of receipt of any requests for convening a conference or other notices.

6. This Treaty shall be registered by the Depositary Governments pursuant to Article 102 of the Charter of the United Nations.

Article X

1. Each Party shall in exercising its national sovereignty have the right to withdraw from the Treaty if it decides that extraordinary events, related to the subject matter of this Treaty, have jeopardized the supreme interests of its country. It shall give notice of such withdrawal to all other Parties to the Treaty and to the United Nations Security Council three months in advance. Such notice shall include a statement of the extraordinary events it regards as having jeopardized its supreme interests.

2. Twenty-five years after the entry into force of the Treaty, a conference shall be convened to decide whether the Treaty shall continue in force indefinitely, or shall be extended for an additional fixed period or periods. This decision shall be taken by a majority

of the Parties to the Treaty.

IN WITNESS WHEREOF the undersigned, duly authorized, have signed this Treaty.

DONE in triplicate, at the cities of London, Moscow and Washington, the first day of July, one thousand nine hundred and sixty-eight.

Entered into force March 5, 1970

In accordance with Article X, paragraph 2, the Review and Extension Conference of the Parties to the Treaty on the Non-Proliferation of Nuclear Weapons decided that the Treaty should continue in force indefinitely on May 11, 1995.

ABOUT THE CONTRIBUTORS

Jeffrey M. Kaplow is an Assistant Professor of Government at the College of William & Mary, where his research focuses on nuclear proliferation, international and civil conflict, and international security institutions. Before coming to William & Mary, he was a fellow with the University of California's Institute on Global Conflict and Cooperation and a Stanton Nuclear Security Fellow at the RAND Corporation. He is the coauthor of the RAND publication, "The Days After a Deal with Iran: Implications for the Nuclear Nonproliferation Regime." In previous work, he analyzed foreign nuclear programs for the U.S. government. He holds a Ph.D. in political science from the University of California, San Diego, a master's degree in international security policy from Harvard's Kennedy School, and a B.A. in political science from Yale.

Victor Gilinsky is an independent consultant primarily on matters related to nuclear energy. He was a two-term commissioner of the U.S. Nuclear Regulatory Commission from 1975-1984, and before that Head of the Rand Corporation Physical Sciences Department. He holds a Bachelors of Engineering Physics degree from Cornell University and a Ph.D. in Physics from the California Institute of Technology, which gave him its Distinguished Alumni Award. He is a member of the American Physical Society and the Institute of Electrical and Electronics Engineers.

Dean Rust served 35 years in the U.S. Arms Control and Disarmament Agency (ACDA) and with the U.S. Department of State. He started in 1970, and after six years of staff work, including a stint in the Office of the ACDA Director, he transferred to the ACDA bureau responsible for nuclear nonproliferation. From then until retirement in 2005, he stayed in the nonproliferation field. He served as Acting Director and Deputy Director of a 10-15 person office on several occasions. Among his primary responsibilities were nuclear export controls including the Nuclear Suppliers Group, civil nuclear cooperation ("123") agreements and the Nuclear Non-Proliferation Treaty. He continued work in this area in the Nonproliferation Bureau of the State Department when ACDA was abolished in 1999. He attended the National War College in 1987-88. For four years after retirement, he consulted with the Los Alamos and Brookhaven National Laboratories.

Henry D. Sokolski is the Executive Director of the Nonproliferation Policy Education Center (NPEC). He previously served as Deputy for Nonproliferation Policy in the Department of Defense, and has worked in the Office of the Secretary of Defense's Office of Net Assessment, as a consultant to the National Intelligence Council, and as a member of the Central Intelligence Agency's Senior Advisory Group. In the U.S. Senate, Mr. Sokolski served as a special assistant on nuclear energy matters to Senator Gordon Humphrey (R-NH) and as a legislative military aide to Dan Quayle (R-IN). He was appointed by Congress to serve on both the Commission on the Prevention of Weapons of Mass Destruction Proliferation and Terrorism in 2008 and the Deutch WMD Proliferation Commission in 1999. Mr. Sokolski has authored and edited a number of works on proliferation, including *Underestimated: Our Not So Peaceful Nuclear Future* (Strategic Studies Institute, 2016) and *Best of Intentions: America's Campaign Against Strategic Weapons Proliferation* (Praeger, 2001).

Leonard Weiss is an affiliated scholar at Stanford University's Center for International Security and Cooperation and a national advisory board member of the Center for Arms Control and Non-Proliferation in Washington, DC. He worked for over two decades for Senator John Glenn as the staff director of both the Senate Subcommittee on Energy and Nuclear Proliferation and the Committee on Governmental Affairs. Dr. Weiss was the chief architect of the Nuclear Nonproliferation Act of 1978 and the legislation creating the Defense Facilities Nuclear Safety Board. He led investigations of the Indian and Pakistani nuclear programs, nuclear security violations at the U.S. weapons labs, and health and safety problems at the Fernald Materials Production Center. Dr. Weiss has published numerous articles for the *Bulletin of the Atomic Scientists*, *Arms Control Today*, and *The Nonproliferation Review*. He has been a consultant to the Lawrence Livermore National Laboratory and the Naval Research Laboratory, and has held tenured professorships at Brown University and the University of Maryland in Applied Mathematics and Electrical Engineering.

Made in the USA
Middletown, DE
14 February 2017